Center for Chemical Process Safety Asia-Pacific Conference 2015

Perth, Australia
9-11 February 2015

ISBN: 978-1-5108-1550-6

Printed from e-media with permission by:

Curran Associates, Inc.
57 Morehouse Lane
Red Hook, NY 12571

Some format issues inherent in the e-media version may also appear in this print version.

Printed by Curran Associates, Inc. (2015)

For permission requests, please contact AIChE
at the address below.

AIChE
120 Wall Street, FL 23
New York, NY 10005-4020

Phone: (800) 242-4363
Fax: (203) 775-5177

www.aiche.org

Additional copies of this publication are available from:

Curran Associates, Inc.
57 Morehouse Lane
Red Hook, NY 12571 USA
Phone: 845-758-0400
Fax: 845-758-2634
Email: curran@proceedings.com
Web: www.proceedings.com

TABLE OF CONTENTS

Shakeel Kadri

- Became CCPS Executive Director on 1 Feb 15 when Scott Berger retired
- At Air Products since 1978 [36 and a half years]
- Held last position as Director of Global Process Safety and Risk Management since 2002
- Have been involved with CCPS since 2002; Actively involved with CCPS activities, with a special focus on risk-based process safety, developing a process safety culture within companies and Metrics.
- On the process safety committees of the American Chemistry Council, the American Petroleum Institute, the Mary Kay O'Connor Process Safety Center, the Compressed Gas Association, the European Industrial Gases and Association, and the American Fuel and Petrochemical Manufacturers Association.
- Fellow of both CCPS and AIChE
- BS and MS in Chemical Engineering and MBA in Finance

The Global Community Committed to Process Safety

1

About CCPS

Not for Profit, Corporate supported – more than 180 members

Vision

In order to protect people, property and the environment, CCPS brings the best process safety knowledge and practices to industry, academia, governments and the public around the world through collective wisdom, tools, training and expertise.

2

An AIChE Technology Alliance
CCPS
Center for Chemical Process Safety

CCPS Mission

Eliminate process safety incidents, in industries, by:

- Promoting process safety as key societal value & expectation.
- Establishing process safety as foundation for responsible operations.
- Serving as the premier world-wide resource for process safety & development of "state-of-the-art" solutions.
- Fostering knowledge & implementation of process safety.
- Advancing process safety education, technology, culture & management practices.

3

An AIChE Technology Alliance
CCPS
Center for Chemical Process Safety

Leading Process Safety since 1985

Creating Books and Publications

Sharing Best Practices

Creating Industry-wide Tools, Programs and Guidelines

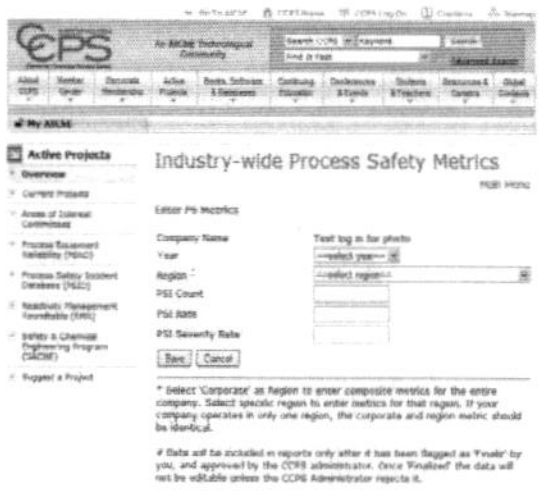

Process Safety Beacon

Conducting Global Conferences and Training

4

Improving safety and environmental outcomes through data analysis – a shared responsibility

Stuart Smith

AIChE-CCPS Asia Pacific Conference
Tuesday, 10 February 2015

NOPSEMA

nopsema.gov.au

NOPSEMA

Notifiable incidents

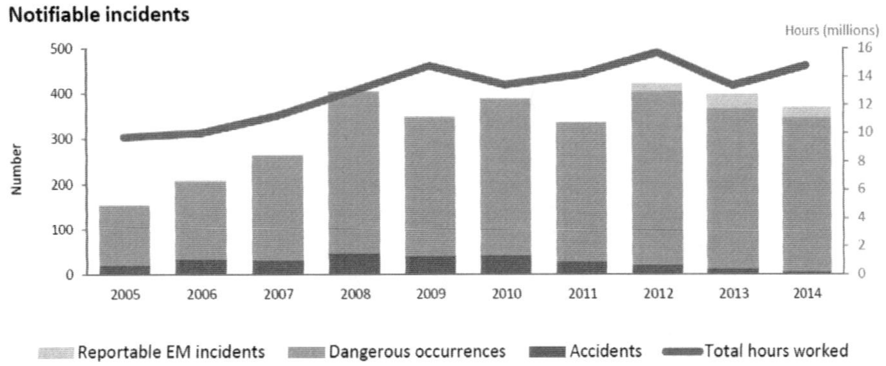

2

NOPSEMA

Dangerous occurrences

Dangerous occurrences - unplanned events, SCE* damage, other

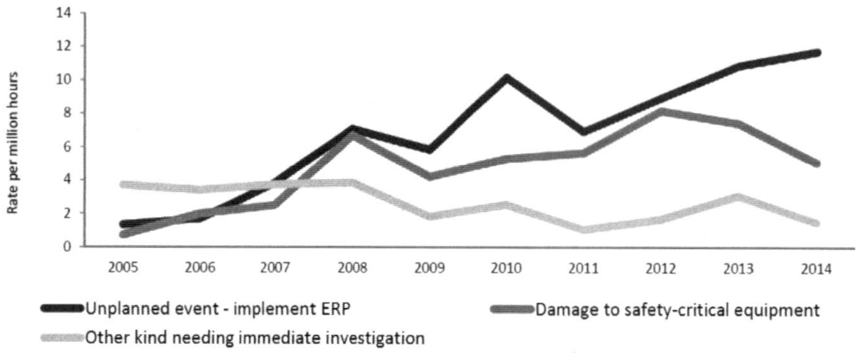

Note : SCE refers to safety-critical equipment.

3

NOPSEMA

Root causes of OHS reportable incidents

2014 OHS incidents - reported root causes

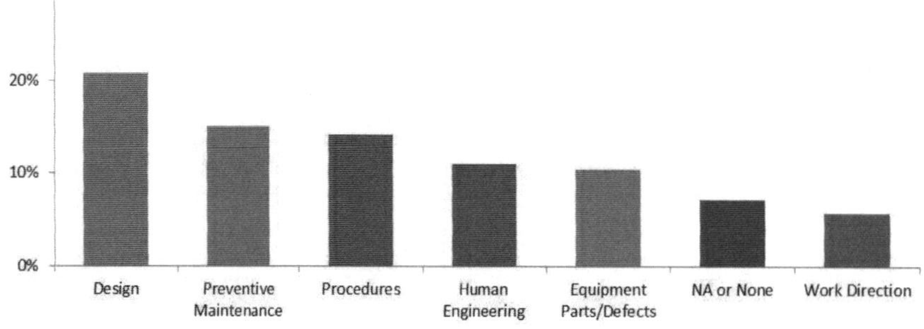

4

NOPSEMA

Hydrocarbon releases

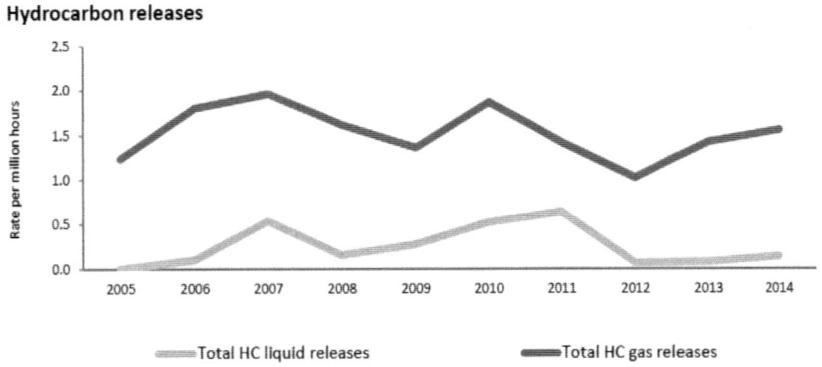

5

NOPSEMA

International benchmarking- Hydrocarbon gas releases

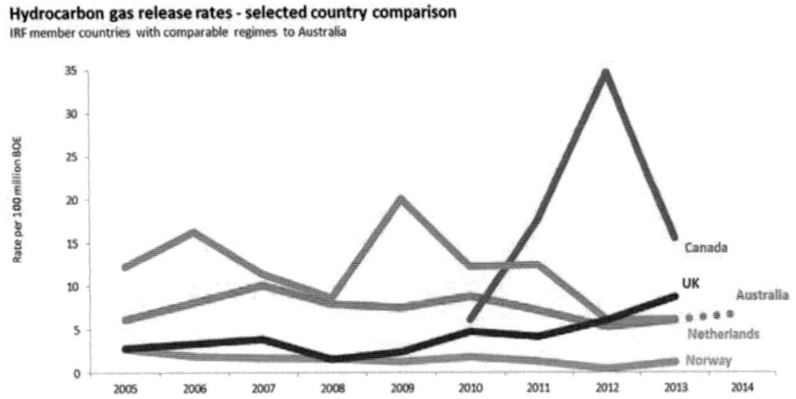

6

NOPSEMA

Injuries

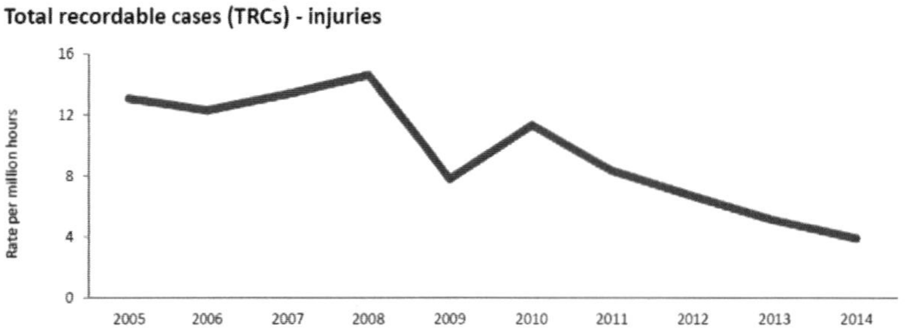

Note: Total recordable cases is the sum of fatalities, major injuries, LTIs (lost time injuries), ADIs (alternative duties injuries) and MTIs (medical treatment injuries)

7

NOPSEMA

International benchmarking-Injury rates

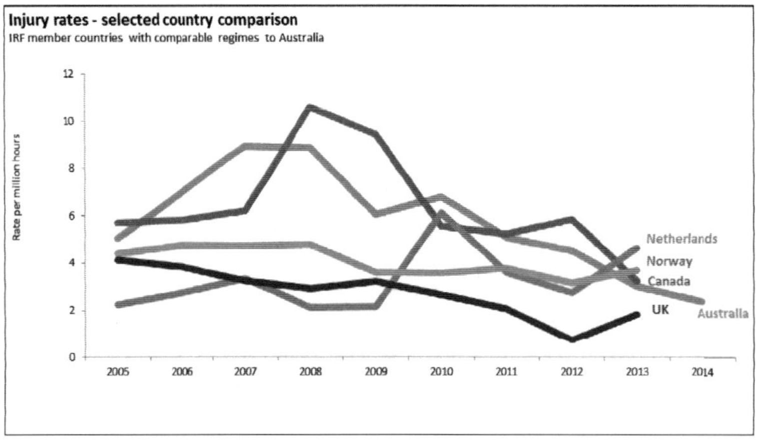

8

NOPSEMA

Assessment and inspection

Assessments notified

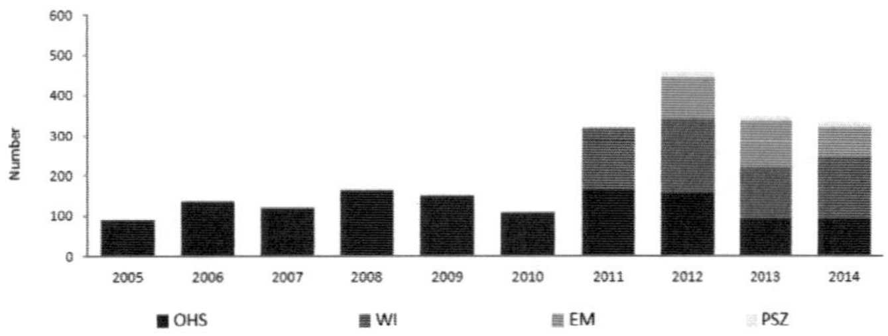

Note: Assessments where NOPSEMA has formally notified the duty holder in writing of a decision.
Only includes assessments that have been notified and completed (i.e. excludes those in progress and recalled/cancelled)

Inspections

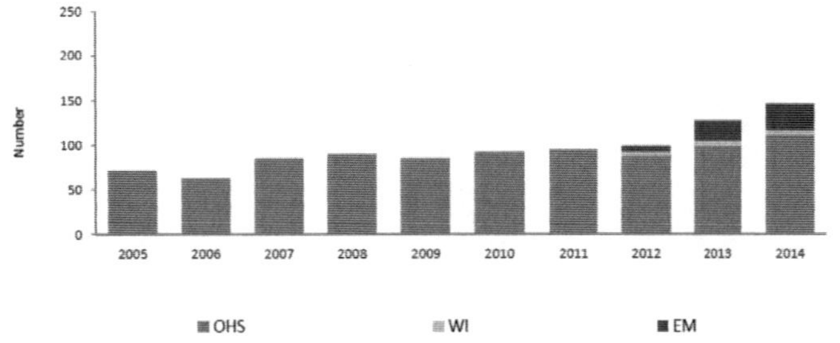

9

NOPSEMA

Enforcement

Enforcement actions

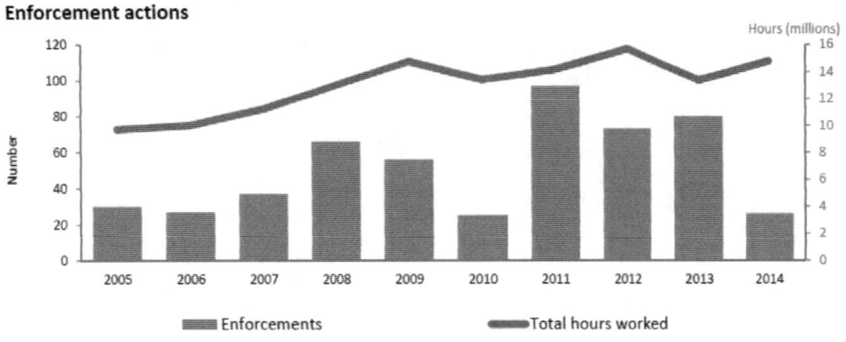

Note: Excludes verbal warnings/advice, directions and investigation notices

10

NOPSEMA

Areas for continued focus

- Performance standards
- Management of change
- Emergency response arrangements
- Blow Out Preventers and associated well control equipment
- Consultation

11

Any questions?

NOPSEMA

nopsema.gov.au

Demonstrating Operational FLNG Safety

woodside

Eloise Browne-Cooper

Risk & Safety Team Lead

10 February 2015

Disclaimer and important notice

This presentation contains forward looking statements that are subject to risk factors associated with oil and gas businesses. It is believed that the expectations reflected in these statements are reasonable but they may be affected by a variety of variables and changes in underlying assumptions which could cause actual results or trends to differ materially, including but not limited to: price fluctuations, actual demand, currency fluctuations, drilling and production results, reserve estimates, loss of market, industry competition, environmental risks, physical risks, legislative, fiscal and regulatory developments, economic and financial market conditions in various countries and regions, political risks, project delay or advancement, approvals and cost estimates.

All references to dollars, cents or $ in this presentation are to US currency, unless otherwise stated.

References to "Woodside" may be references to Woodside Petroleum Ltd. or its applicable subsidiaries.

2

Agenda

Demonstrate how Woodside will safely operate Floating Liquefied Natural Gas (FLNG) facilities through the application of:

- Woodside's floating facility and LNG processing experience
- Woodside's safety management system
- Woodside's risk management process

3

Woodside's operating facilities

4

Woodside's operational experience

- Operate
  - 4 Floating Production and Offloading Facilities (FPSOs)
  - 4 fixed offshore facilities
  - 5 LNG trains at Karratha Gas Plant
  - 1 LNG train at Pluto LNG
- Pluto Support Centre – Similar remote operations philosophy to FLNG
- Internal Floating LNG technology
- WITH: Woodside Innovation and Technology Hub

5

Combining LNG and floating experience - FLNG

6

Woodside Management System

wms

Compass and Policies ①	Values, policies, mission, strategic direction
Expectations ①	Essential activities to achieve business objectives
Procedures ⑯	How to apply processes & activities
Guidelines ㉛	Required controls, recommended practices

Woodside risk management process

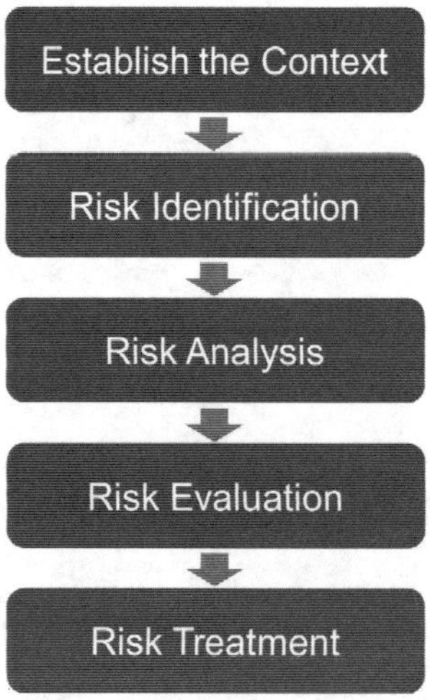

Hierarchy of risk management controls

Decreasing effectiveness

- Elimination of risk by removing the hazard
- Substitution of a hazard with a less hazardous one
- Prevention of potential events
- Separation of people from the consequences of potential events
- Control of the magnitude and frequency of an event
- Mitigation of the impact of an event on people
- Emergency response and contingency planning

9

Inherently safer design goals

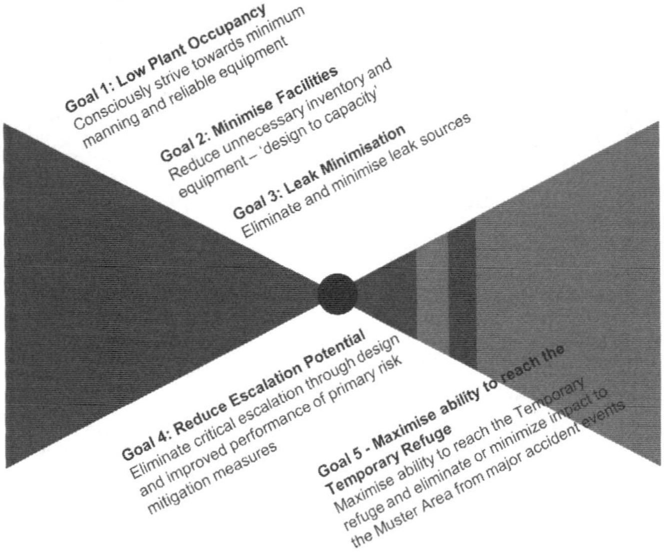

10

FLNG safety in design

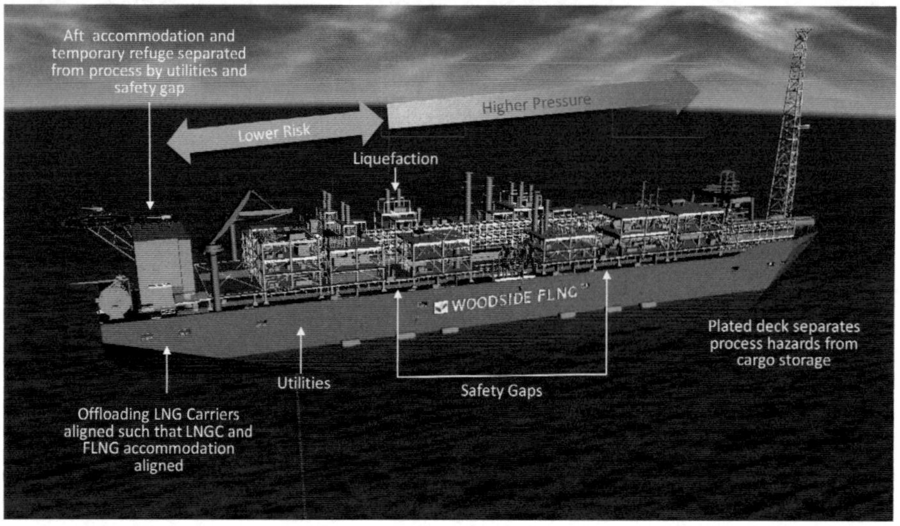

11

Formal safety assessment

- Hazard identification, including Major Accident Events (MAEs)
- Fire and explosions assessment
- Assessment of other MAEs (dropped objects, ship collision risk and structural integrity)
- Escape, temporary refuge, evacuation and rescue provisions, including emergency response arrangements
- Assessment of key control measures and development of associated performance standards
- Demonstration that risks are tolerable and as low as reasonably practicable

HAZARD IDENTIFICATION
- HAZID/HAZOP
- MAEs

RISK ANALYSIS
- MAE Consequence
- Flammable Hazard Analysis
- Non-Flammable Hazard Analysis
- Escape, Temporary Refuge, Evacuation & Rescue Analysis
- Safety Critical Element Analysis
- Quantitative Risk Assessment

ID OF TECHNICAL CONTROLS
- Safety critical Elements
- Performance Standards

ALARP Demonstration

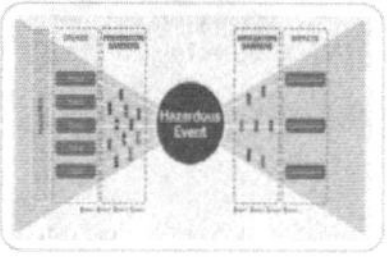

12

Typical FLNG Major Accident Events

- Typical FLNG MAEs are similar to FPSO MAEs.
- Managed by the same Formal Safety Assessment process.
- The following MAEs have specific FLNG differences:
  - Flammable hazards
  - Ship Collision
  - Structural Integrity
  - Loss of Position or Stability

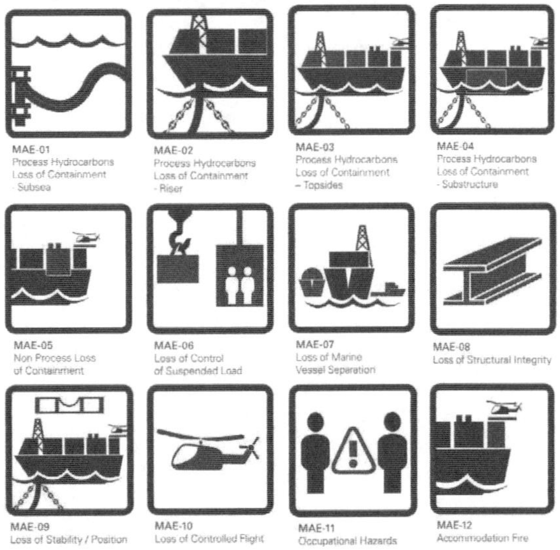

13

FLNG flammable hazard risk management

- Minimisation of leak sources through the reduction in unnecessary equipment, connections and flange joints
- Control of ignition sources through hazardous area classification and layout
- Maximise ventilation of process areas to reduce build-up of flammable gas
- LNG and condensate storage tanks operate at atmospheric pressure
- Self-draining decks to remove liquid hydrocarbon releases as quickly as possible
- Application of offshore philosophy to emergency shutdown and depressurisation
- Passive fire protection and cold spill protection
- Separation and segregation of temporary refuges from flammable hazards

14

FLNG ship collision risk management

- Hull design to withstand low energy collisions from LNG carrier and condensate tanker
- Substructure double skinned over full length of LNG and condensate tanks
- Fenders protect structure from contact by visiting LNG carrier
- Facility based navigation aids, collision avoidance systems, radar and communication systems
- Adequate space for offtake carriers to make approach
- Turret mooring system and risers protected from external impact events.

15

FLNG structural integrity management

- Designed to withstand most onerous environmental conditions
  - 10,000 year return period extreme weather event
  - Category 5+ cyclone

16

FLNG loss of position or stability

- Substructure designed to relevant International Maritime Organisation (IMO) requirements
- Redundancy in mooring line and riser design
- Ballast arrangement providing sufficient immersion of the hull in severe weather conditions
- FLNG has positive stability in all conditions
- Mooring Load Monitoring system used to constantly monitor tension in the mooring lines

17

Conclusion

- Over 25 years of safe operational experience which is applicable to the safe operation of FLNG
- Demonstrate operational safety through application of the Woodside Safety Management System
- FLNG risks managed as per all offshore and LNG facilities
- Risks reduced to tolerable and as low as reasonably practicable.

Image courtesy of Shell

18

WOOD GROUP KENNY

Does FLNG raise new safety issues?

Dr. James Holbeach
SURFA and Process Simulation Manager
Wood Group Kenny Australia

e:james.holbeach@woodgroupkenny.com
p:+61 417 553 757

AIChE CCPS 2015 Asia Pacific Conference, Feb 10th 2015, Parmelia Hilton, Perth Australia

Safety & Assurance | Relationships | Social Responsibility | People | Innovation | Financial Responsibility | Integrity

FLNG Context

Typical Traditional Offshore LNG

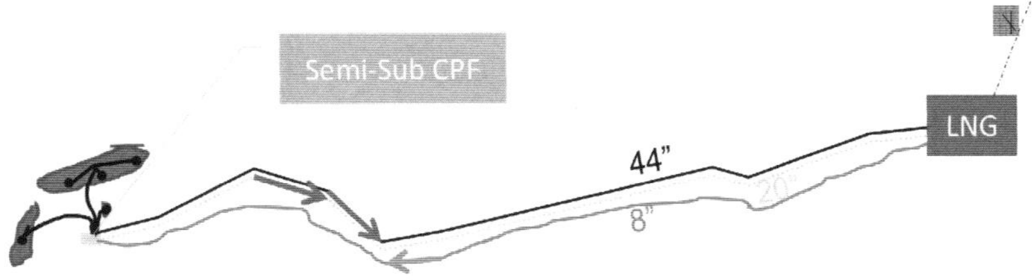

Central Processing Facility (CPF)
- Subsea gathering / control
- Dehydration
- Partial condensate stabilisation
- Water disposal
- Compression

Onshore LNG Plant
- Airport / Roads / Infrastructure
- Processing / Liquefaction
- Offloading (LNG / Condensate etc.)
- Chemical Regeneration
- Loading (Chemicals)

Graphic for illustrative purposes only. Not based on any facility, real or proposed

Floating LNG

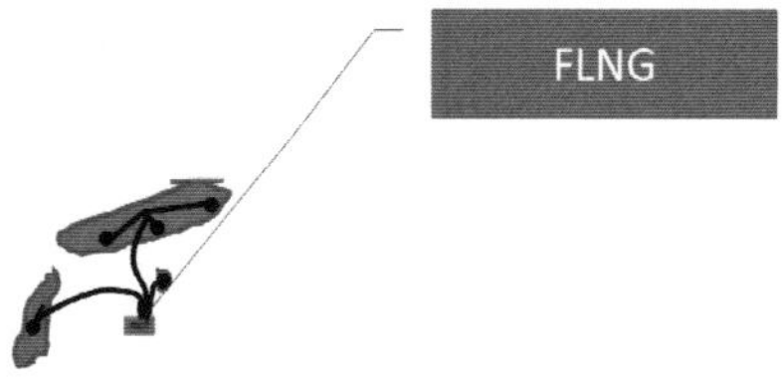

Floating LNG Facility (FLNG)

- **Subsea gathering / control**
- **Dehydration**
- **Condensate stabilisation**
- **Water disposal**
- **Processing / Liquefaction**
- **Offloading (LNG / Condensate etc.)**
- **Chemical Regeneration**
- **Loading (Chemicals)**

Graphic for illustrative purposes only. Not based on any facility, real or proposed

Control

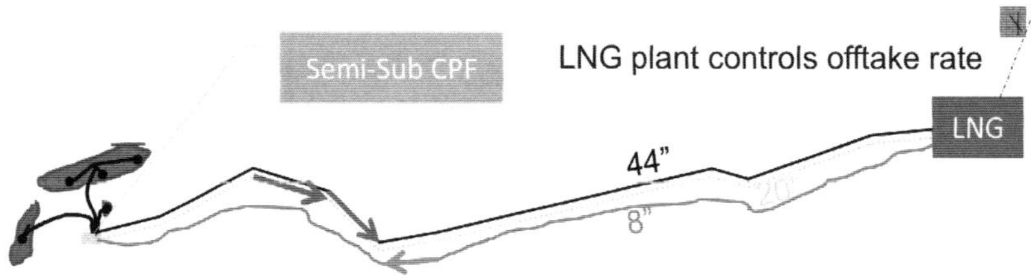

Wellhead chokes control overall flowrate to LNG Facility

Graphic for illustrative purposes only. Not based on any facility, real or proposed

Weather Forecasters

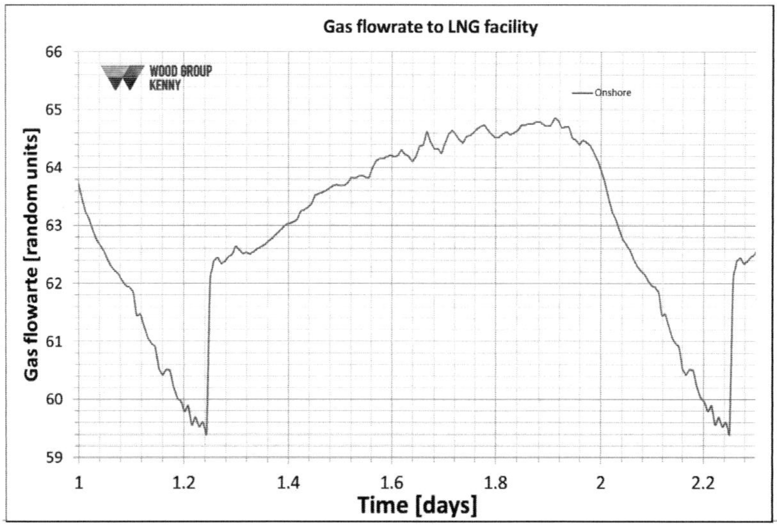

Production is never simple

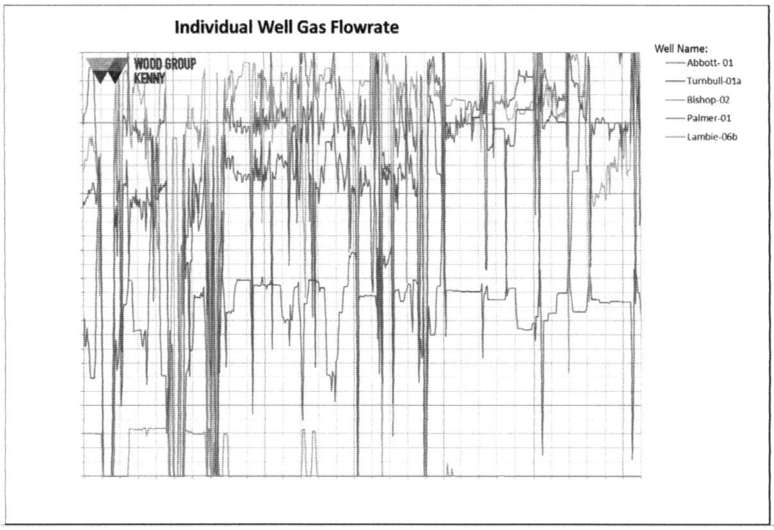

The Biggest Difference

What is the biggest difference between a traditional LNG and Floating LNG?

Project	Type	Pipe length	Diameter	Volume* [m³]	Size [MTpa]	Volume to Size [m³/Mtpa]
Gorgon / Jansz	Wet Gas	70 km 70 / 80km	34" 30 / 34"	114,000	15	7600
Pluto	Dry Gas	180 km	36"	118,000	4.3	27500
Wheatstone	Dry Gas	220 km	44"	216,000	8.9	24200
NW Shelf	Dry gas	135 km	40/42"	230,000	16.3	14000
GLNG	Dry Gas	420 km	42"	375,000	7.8	48100
Ichthys	Dry Gas	889 km	42"	795,000	8.4	94600
FLNG	**Wet Gas**	**3 x 6 km**	**12"**	**1,300**	**6**	200

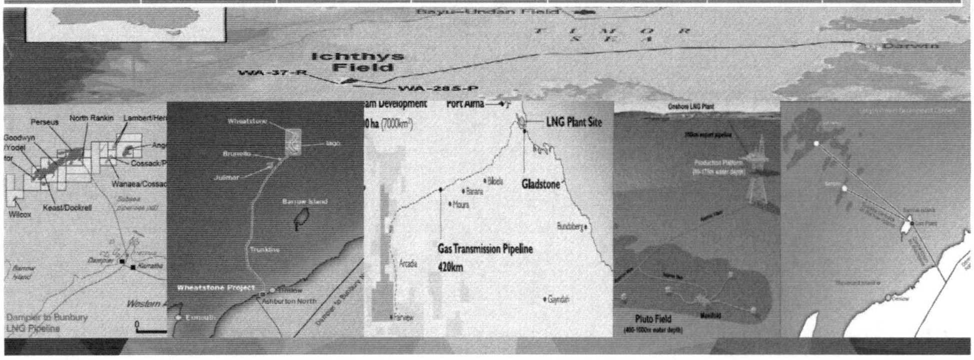

Parameter	Low	High	Units
Flowrate [BScfd]	1.1	1.5	BScfd
Pressure (Arrival)	60	65	Bara
Temp (Arrival)	10	15	°C
Ullage (75%)	120	600	min
Restart Time	0	360	min

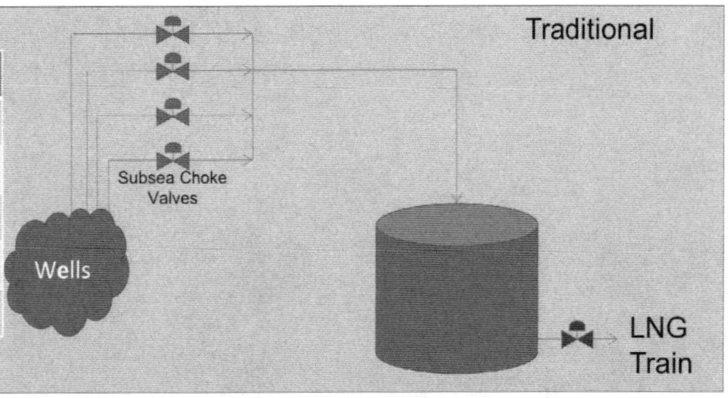

Parameter	Low	High	Units
Flowrate [BScfd]	1.1	1.5	BScfd
Pressure (Arrival)	60	95	Bara
Temp (Arrival)	-20	120	°C
Ullage (75%)	0	10	min
Restart Time	0	360	min

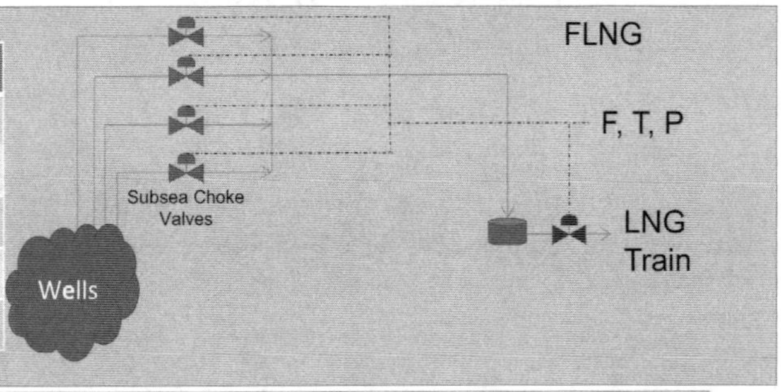

Other Issues

Process FLNG:

- Black Start-up
- Temperature / Liquid surge
- Fatigue / reliability
- Instrumentation criticality
- Potential Flaring
- New unproven equipment

Financial Drivers (All LNG)

- Cargo Schedule
- Deferred production (NPV)
- Cost recovery
- Reputation
- Loss of spotprice trading
- Loss of product (flaring)

= Don't Shut down
Don't ramp down
Keep producing

13 - Wood Group Kenny

Assumptions ...

"Questions are the antidote to assumptions that so often incubate mistakes"

Sir Charles Haddon- APPEA, Perth 2014

14 - Wood Group Kenny

Do these differences in upstream impact FLNG safety?

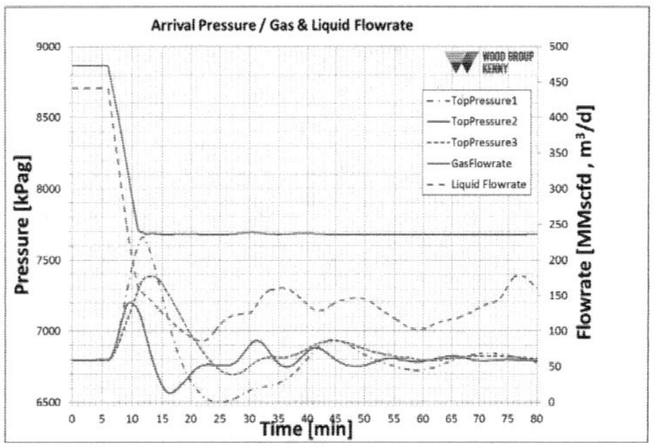

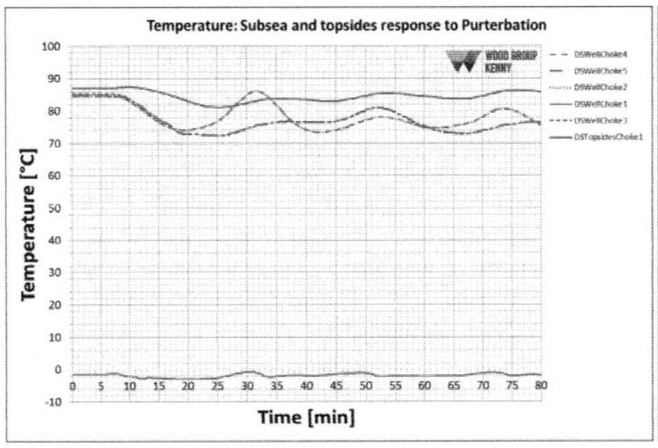

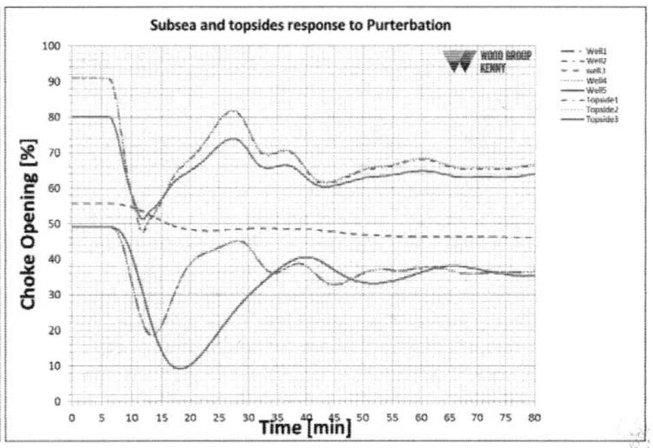

WOOD GROUP KENNY

Safety & Assurance | Relationships | Social Responsibility | People | Innovation | Financial Responsibility | Integrity

DNV·GL

Key Safety Challenges for FLNG

Darren McFarlane – DNV GL

AIChE-CCPS Asia-Pacific Conference – Perth Feb 2015

Presentation Content

- **DNV GL in Brief**
- **FLNG Options**
- **Some Safety Issues**
- **Conclusions**
- **Questions**

Industry consolidation

DNV·GL

DNV | GL Group | KEMA | NOBLE DENTON | GARRAD HASSAN | ADVANTICA

Detailed Service Overview

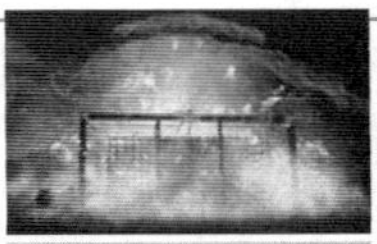

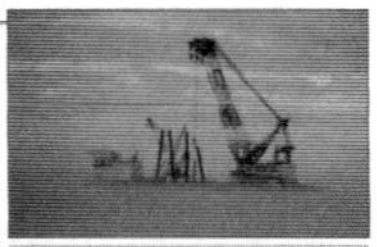

Technical Assurance	Risk Management Advisory	Technical Advisory – Asset Integrity	Technical Advisory – Process & Engineering	Noble Denton Marine Assurance & Advisory
Design appraisal Certification Verification / Validation Vendor Surveillance Third Party Inspection Expediting Inspection and quality assurance - During EPIC phase (at vendor and site) - During operational phase	Enterprise Risk Management Asset Risk Management Environmental Risk Management Safety Case Management Process Safety HAZOP / HAZID / SIL / FMECA / FS / SCE / PS / QRA / EMERA Bow-tie Fault tree analysis Fire, explosion, ignition & dispersion modelling	Asset integrity management planning Asset life extension Defect analysis Corrosion analysis and mitigation Materials testing and failure analyses Pipeline geotechnics Full scale testing Pipeline engineering SRA	Technical due diligence Gas processing Asset optimisation (RAM) Performance forecasting Rotating machinery Condition & performance monitoring Flow Assurance Feasibility & Pre FEED studies Technical due diligence Gas meter engineering, calibration / validation	Marine warranty Marine consulting Marine casualty investigations Marine operations support Loading and unloading analysis Dynamic positioning

DNV GL has participated in defining the LNG industry

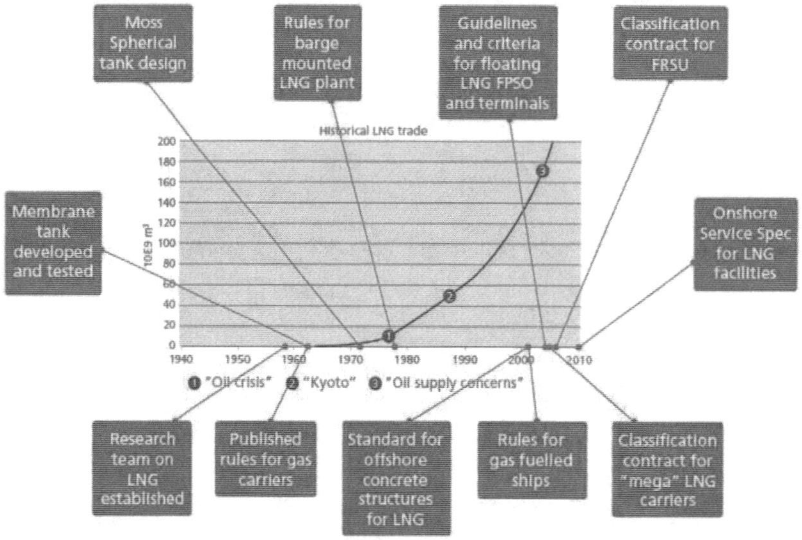

DNV GL classifies a large share of the global floating LNG fleets

	World*	DNV GL
FSRU delivered	7	4
FSRU on order	7	6
RV delivered	7	2
FLNG on order	5	2

FSRU - Floating Storage and Regasification Unit
RV - Regasification vessel
FLNG - Floating Production of LNG

Presentation Content

- DNV GL in Brief
- **FLNG Options**
- Some Safety Issues
- Conclusions
- Questions

FLNG Variants

Shell Prelude

Petronas FLNG1

Golar FLNG

Presentation Content

- DNV GL in Brief
- FLNG Options
- **Some Safety Issues**
- Conclusions
- Questions

Sloshing

reduced magnitude of sloshing effects in lower resonance period range

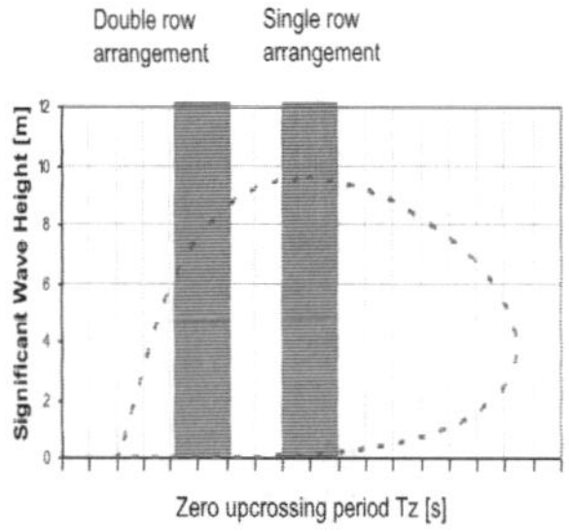

reduced resonance period will reduce probability of resonant wave encounters

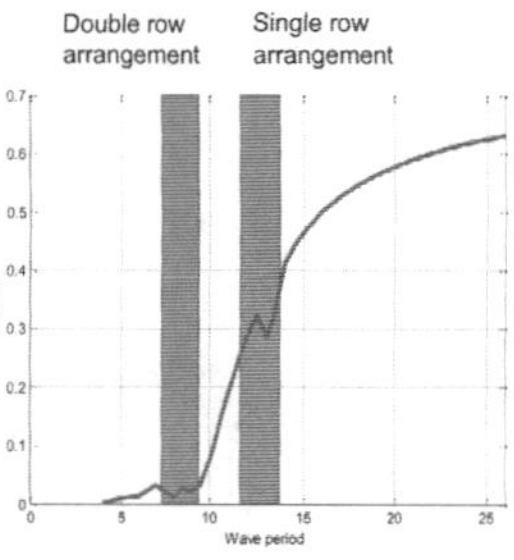

Sway

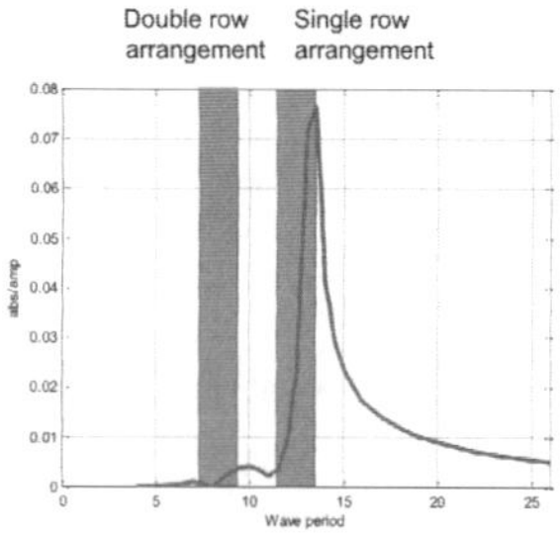

Roll

Different sloshing phenomenon as function of filling

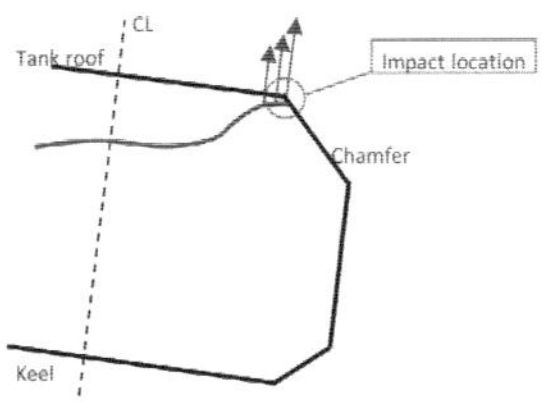

High-filling (~70-100%H) impact due to longitudinal movement

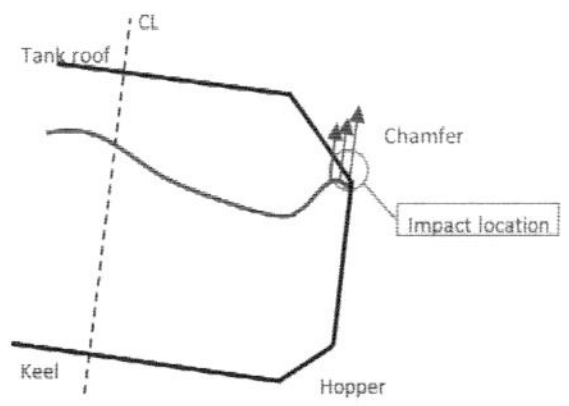

High-filling (~60-70%H) impact due to a run-up against the longitudinal and or transverse bulkhead

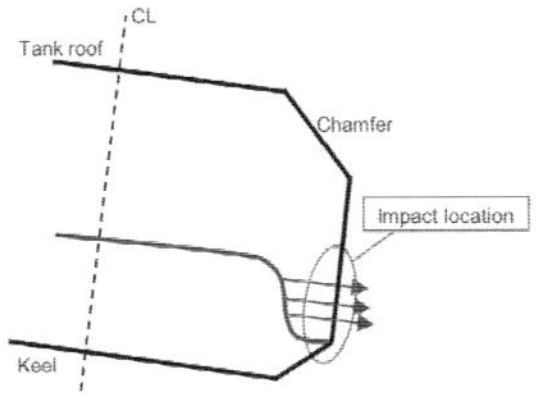

Low-filling (~10-40%)*hydraulic jump*

Light gas leak

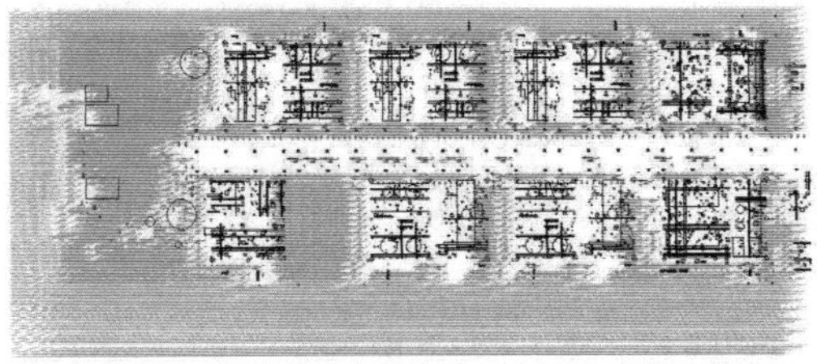

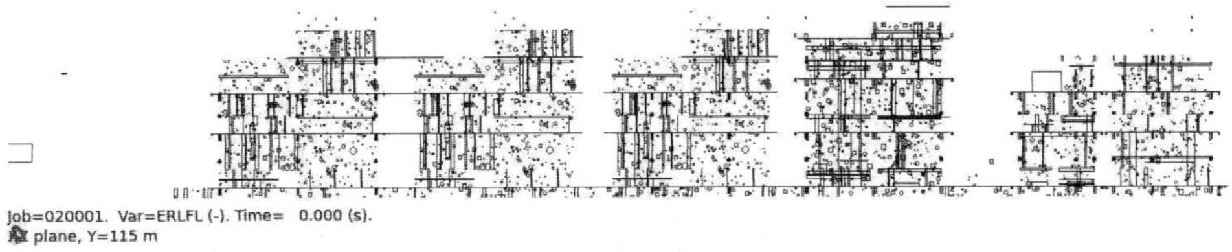

Heavy gas leak

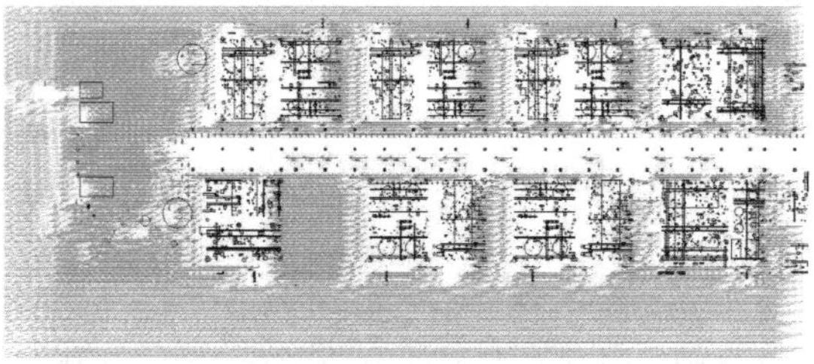

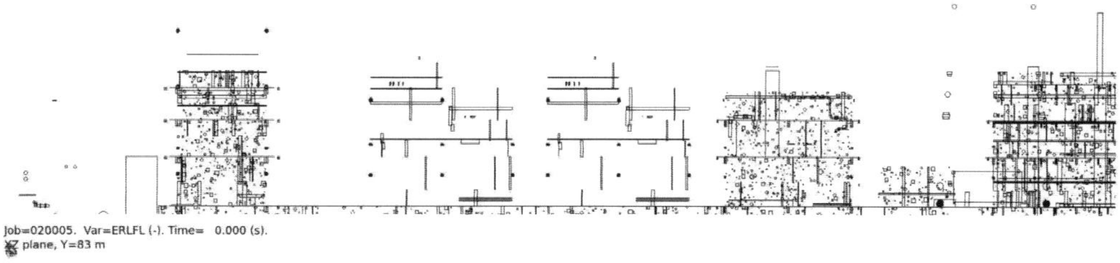

Experiments and simulations

- Major Hazards Research and Testing Facility (Spadeadam)
- Enables us to understand hazards and to develop and validate models

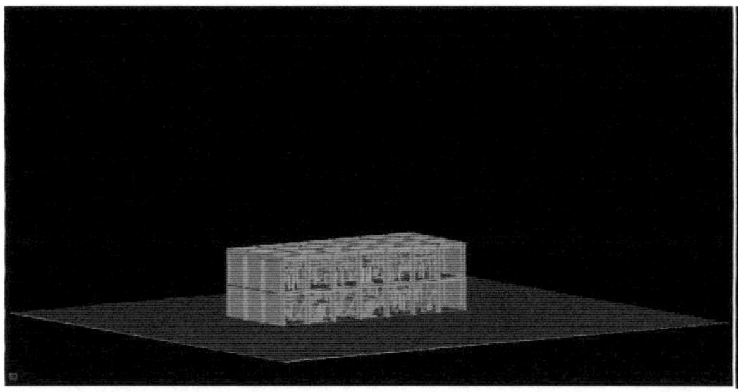

Safety Gaps

Testing at DNV GL Spadeadam

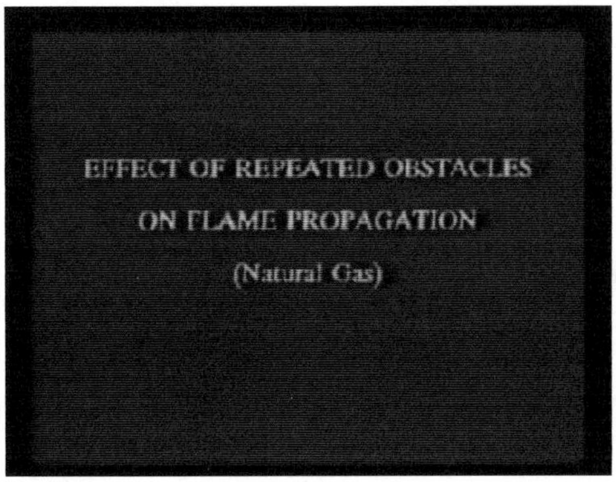

Flame Propagation – filled module

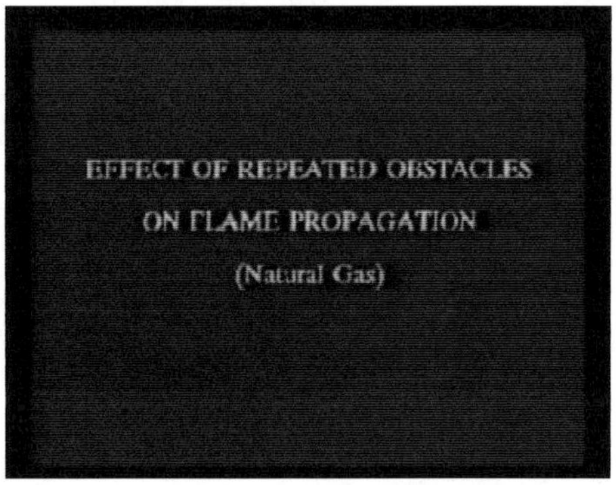

Flame Propagation – simulated gap

Pressure reduction from "safety gaps"

Sett ovenfra t = 1

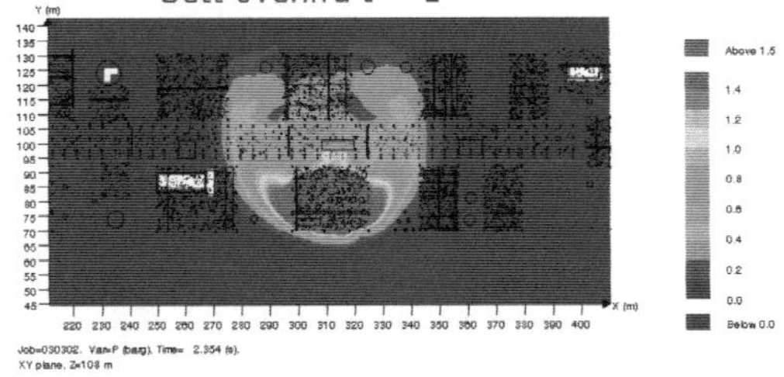

Sett ovenfra t = 2

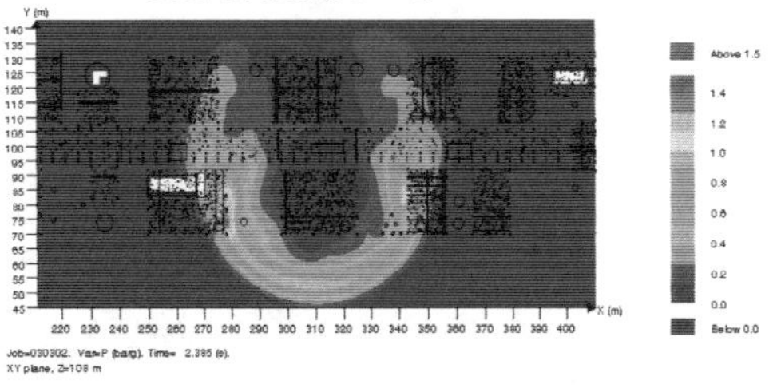

Sett fra siden t = 1

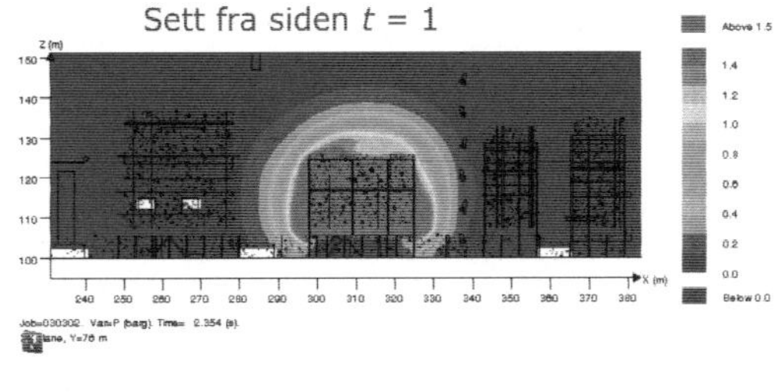

Sett fra siden t = 2

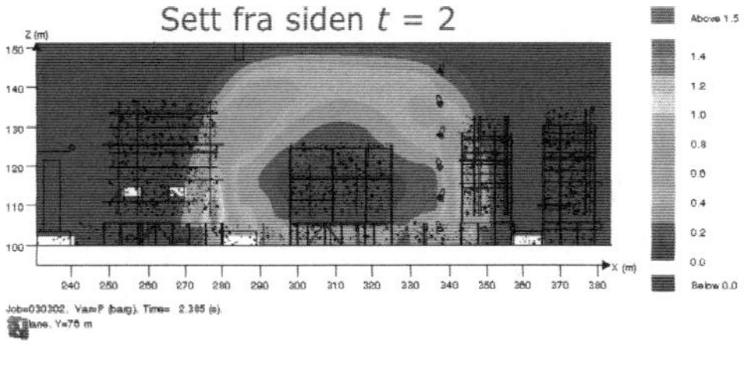

Pick-up after safety gap – heavy gas

Sett ovenfra t = 1

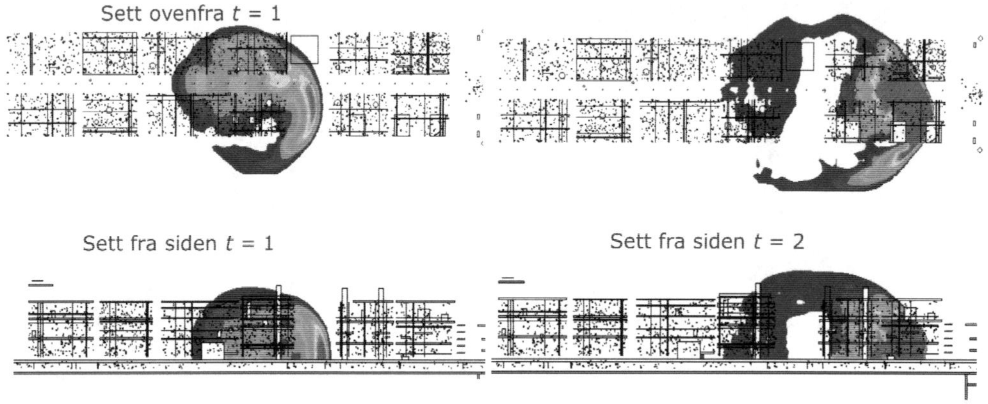

Sett fra siden t = 1

Sett fra siden t = 2

Sett ovenfra t = 2

Dilemmas and their effect on the fire and explosion risk

- Ventilation vs. Working environment
  - Excessive use of wind protection and winterization causes reduced ventilation
  - Optimal ventilation:
    - minimizing wind protection to give acceptable availability
    - Maximizing explosion ventilation to give acceptable explosion risk and DAL pressures
    - A compromise can be found by modelling both using same models.
- Firewalls (relevant for FLNGs, etc.)
  - Stops fire and explosion escalation
  - Reduced ventilation causes small leaks to make large gas clouds,
  - Increase explosion pressure due to reflection wave and larger clouds,
- PFP vs explosion and fire loads
  - Adding insulation on all pipes can cause pressure to increase significantly
  - PFP also increases fire heat due to no heat can be absorbed in structure and piping
  - It is therefore recommended to minimize use of PFP
- PFP vs Flare
  - Flare has more benefits than PFP and should be increased first to max capacity
- Consider also pipe wall thickness as a means of fire protection

Cryogenic Risk and Response Analysis

- Section 1: Risk Analysis to find DAL scenario
- Section 2: Response analysis to optimize Cryogenic protection

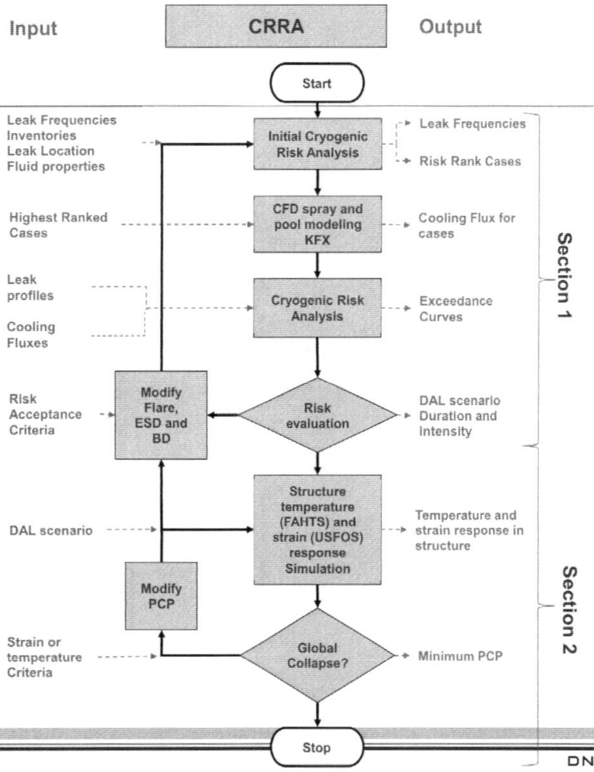

Physical effects considered- CRRA – step by step

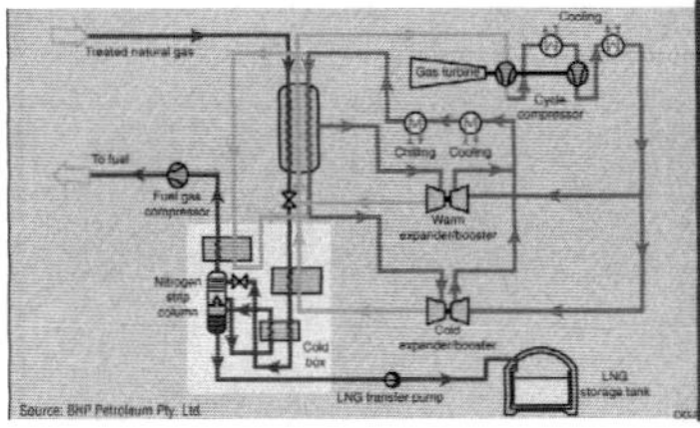

Process conditions; HYSYS Layout

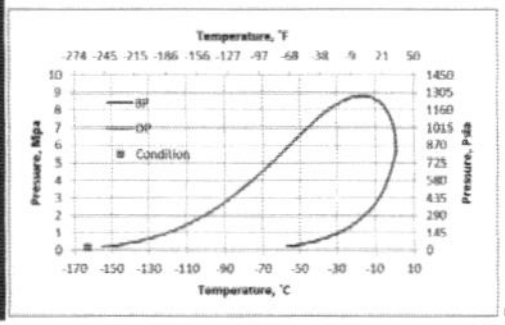

Phase changes during leak

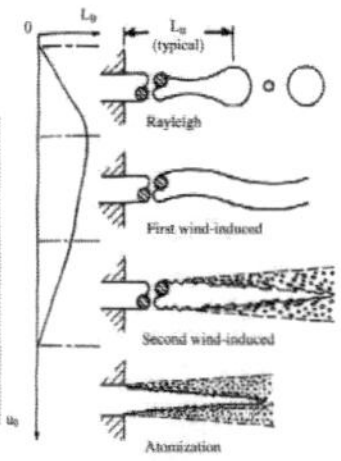

Release conditions

Spray effects, KFX-LNG

Surface heat transfer KFX-LNG and FAHTS

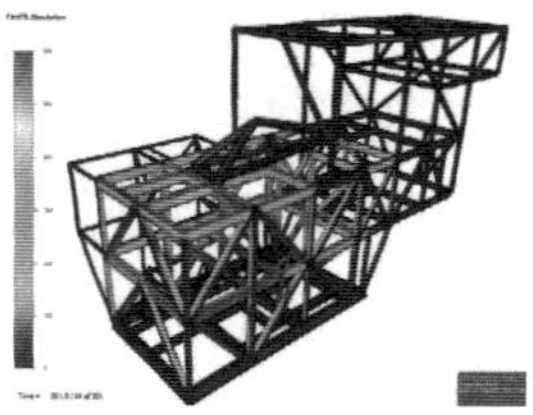

Temperature and strain response FAHTS/USFOS

Presentation Content

- DNV GL in Brief
- FLNG Options
- Some Safety Issues
- **Conclusions**
- Questions

Conclusions

- Sloshing on the LNG carrier is now more important than on the FLNG unit
- Fire and Gas issues are even more important on FLNG compared to FPSO
- Data and models determine accuracy of analyses
- Adjusting design parameters may have conflicting safety effects
- Need for a more accurate cryogenic protection model
- Additional learnings will come once FLNG units become operational

Presentation Content

DNV GL in Brief

FLNG Options

Some Safety Issues

Conclusions

Questions

Questions

Regulatory and Safety Challenges of FLNG

AiChE-CCPS 2015
Darren.Mcfarlane@dnvgl.com
Head of Department - Verification and Risk Advisory
+61 41 774 8883

www.dnvgl.com

SAFER, SMARTER, GREENER

Background slides

Requirements

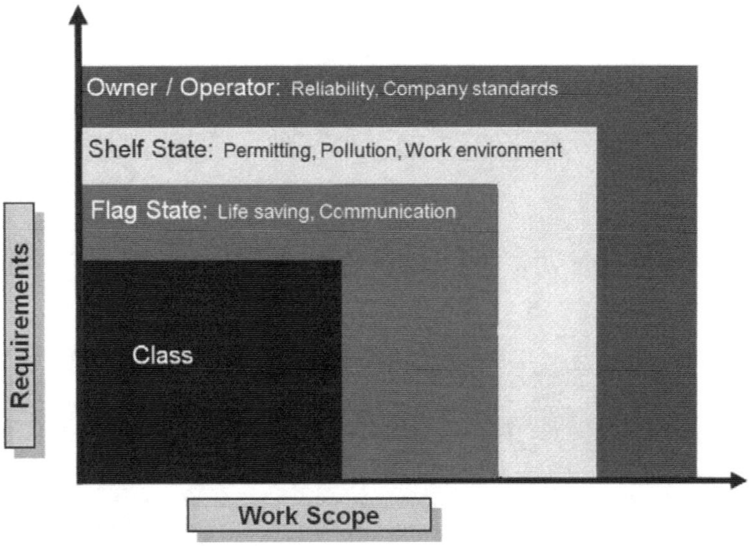

DNV Rules for LNG FPSOs

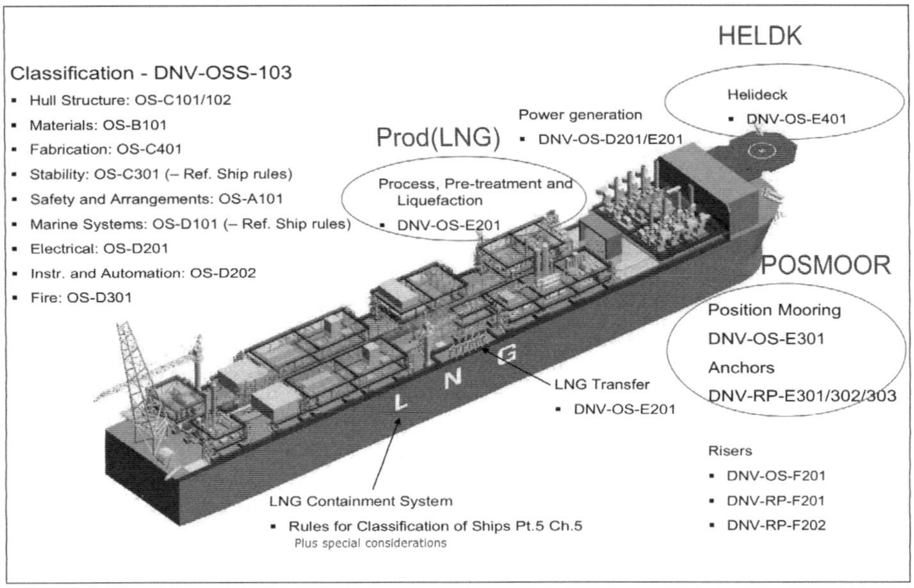

Flag State Requirements (main technical Conventions)

Based on International (IMO) Conventions

- SOLAS (Safety of Life at Sea)
- Load Line
- MARPOL (Marine Pollution)

- IGC Code (Gas Code)
  - "Floating Production, Storage and Offloading (FPSO) facilities, which are designed to handle liquefied gases in bulk, do not fall under the IGC Code. However, designers of such units may consider using the IGC Code to the extent that the Code provides the most appropriate risk mitigation measures for the operations the unit is to perform. Where other more appropriate risk mitigation measures are determined that are contrary to this Code, they shall take precedence over this Code." - proposed IGC Code update

DNV Publications for Classification of LNG FPSOs

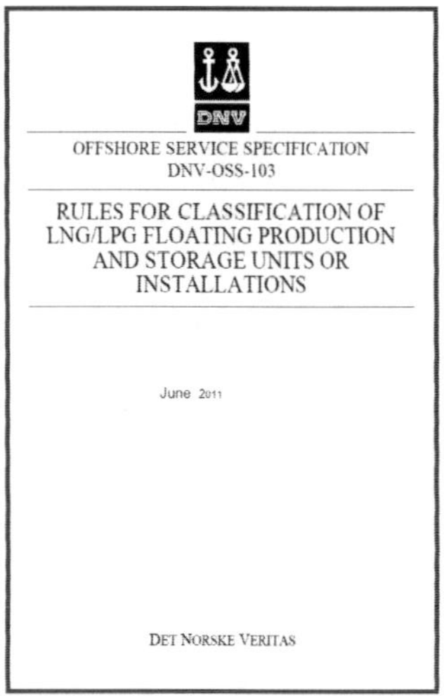
DNV

OFFSHORE SERVICE SPECIFICATION
DNV-OSS-103

RULES FOR CLASSIFICATION OF LNG/LPG FLOATING PRODUCTION AND STORAGE UNITS OR INSTALLATIONS

June 2011

DET NORSKE VERITAS

OSS 103 - Rules for LNG FPSOs

OTG-02 - Guidance on offshore LNG

Safety Case Approach

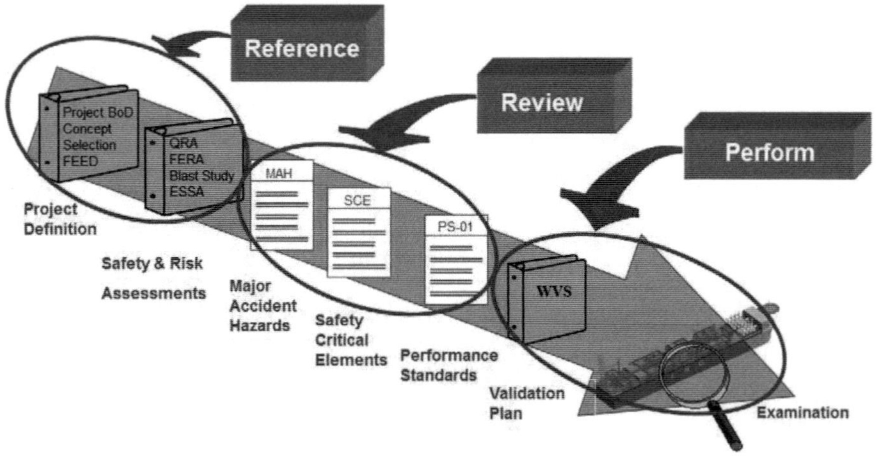

What about novel technology?

DNV RP A203 – Qualification of Technology

Trelleborg Floating Hose

Technip ALLS

Novel technology – manage the risks upfront

- For large scale concepts (i.e.FLNGs)
  - Approval in Principle is typically applied *(OTG-02 Appendix C)*
    - Independent assessment of a concept
    - Confirmation that a design is feasible & no "showstoppers" would prevent the concept being realised.
    - Typically carried out at an early stage of a project
    - Confirms feasibility towards the project team, company management, external investors or future regulators.
    - Based on a limited level of engineering detail, focus on the major hazards to a project.
    - Can be a stand-alone study or a step towards achieving full Classification approval.

- More discrete subsystems and components (i.e. LNG transfer system)
  - Qualification of Technology typically applied (*DNV-RP-A203*)

Trelleborg Floating Hose

33

and organisational

Human factors in safety: the top 10 issues

Ronny Lardner AFIChemE
Registered Psychologist
ronny@keilcentre.com.au

THE BRITISH ASSESSMENT BUREAU ISO9001
The British Psychological Society
Chartered Psychologist
IEHF Registered Consultancy

the Keil centre
chartered psychologists & ergonomists

Overview

- Top-ten "human and organisational factors" most relevant to high-hazard industries
- UK oil and gas industry experience of making improvements on these topics
- A case study
- Further reading & resources
- Questions / comments

STEP CHANGE IN SAFETY

the Keil centre
chartered psychologists & ergonomists

Have you ever done the following....?

- Knowingly broken the speed limit
- Gone into a room, and forgotten why you went there?
- Stolen something
- Not seen something you want in the supermarket, even though its right in front of you
- Done home improvement work, without full PPE
- Daydreamed when you should be paying attention
- Failed to pay your train fare
- Bumped into things or people?
- Told a "white lie"
- Put the right thing in the wrong place (e.g. milk into oven)

Which work behaviours are a problem....?

- Driving over speed limit on company business
- Forgetting an important step in a procedure
- Deliberately taking banned items onto site
- Mis-reading a manual or drawing
- Not using the PPE provided
- Confusing two similar pieces of equipment, or products
- Not wearing seatbelts on fork-lift truck
- Misunderstanding a work instruction
- Using a mobile phone whilst driving
- Opening the wrong valve by mistake

Why "human factors"?

- UK Health and Safety Executive
  - General guidance on human factors HS(G)48
  - 10 key human factors topics for hazardous sites / installations
  - Top-ten adopted / adapted by NOPSEMA, DMP
- BP Texas City
  - Lost balance between process & occupational safety
  - Many human factors causes
  - "behavioural safety" not enough
- Client / partner expectations
  - Developing their own internal expertise
  - e.g. Woodside, Chevron, Inpex

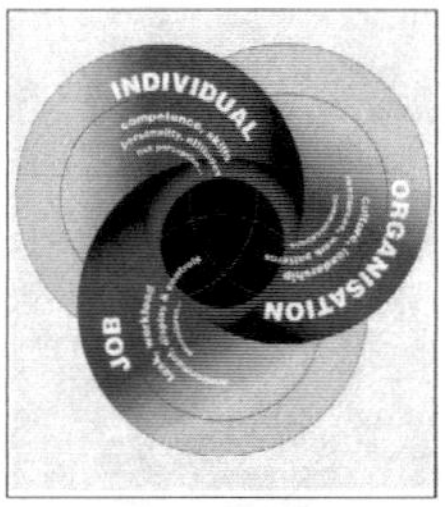

the Keil centre chartered psychologists & ergonomists

Human factors

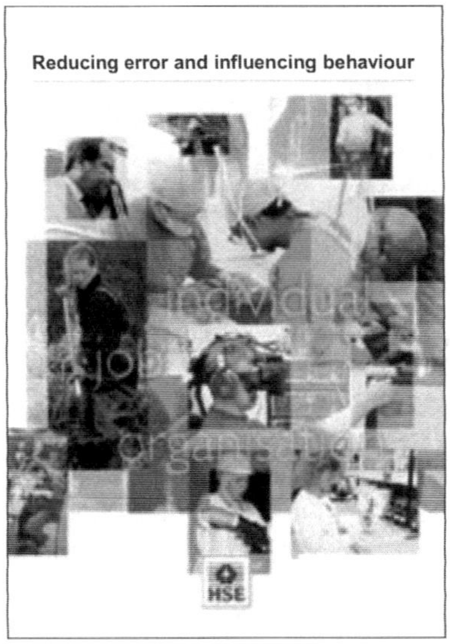

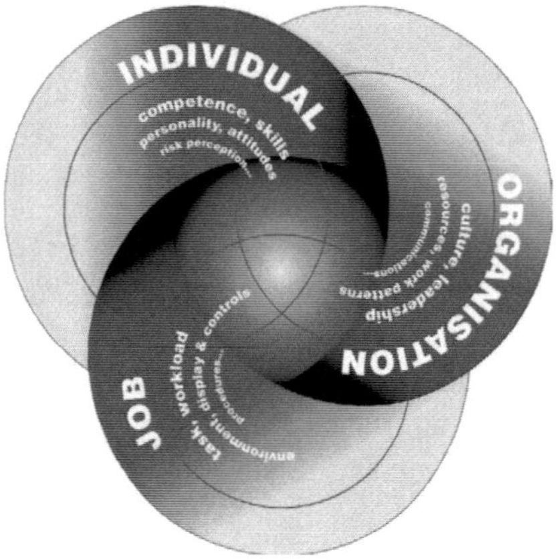

the Keil centre chartered psychologists & ergonomists

7

What do we mean by 'human factors' in hazardous industries?

10 key UK HSE human & organisational factors issues

- Managing human failure
  - During risk assessment
  - As part of incident analysis
- Procedures
- Training and competence
- Staffing, including supervision
- Organisational change
- Safety-critical communication
- Human factors in design
- Fatigue & shiftwork
- Organisational (safety) culture
- Maintenance, inspection and testing

the Keil centre chartered psychologists & ergonomists

8

GOAL: Managing human reliability & failure

↑

Influences on human reliability & performance	
Organisational (safety) culture	
Usable procedures	Safety-critical communication
Training and competence	Human factors in design
Staffing levels, including supervision	Fatigue & shiftwork
Organisational change	Maintenance, inspection and testing

the Keil centre chartered psychologists & ergonomists

9

Case study: effectiveness of post-incident learning

- Serious maritime incident
- This incident involved a very experienced ship's captain
- Initial focus on blame, and immediate causes
- Human factors analysis led to
  - Identification of organisational / system causes
  - Better methods of post-incident learning

the Keil centre chartered psychologists & ergonomists

10

Human Factors Analysis Tools

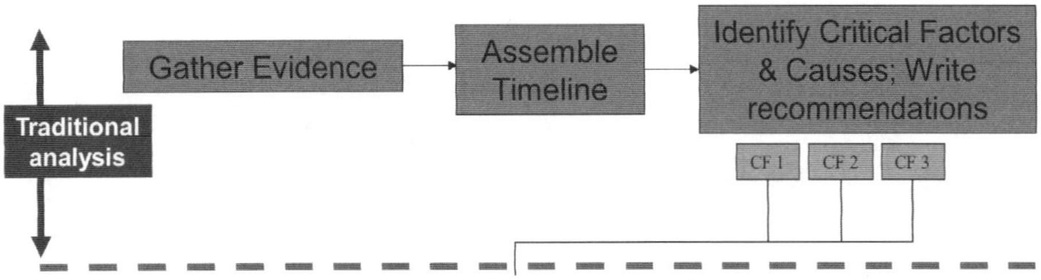

the Keil centre chartered psychologists & ergonomists

Human Factors Analysis Tools

11

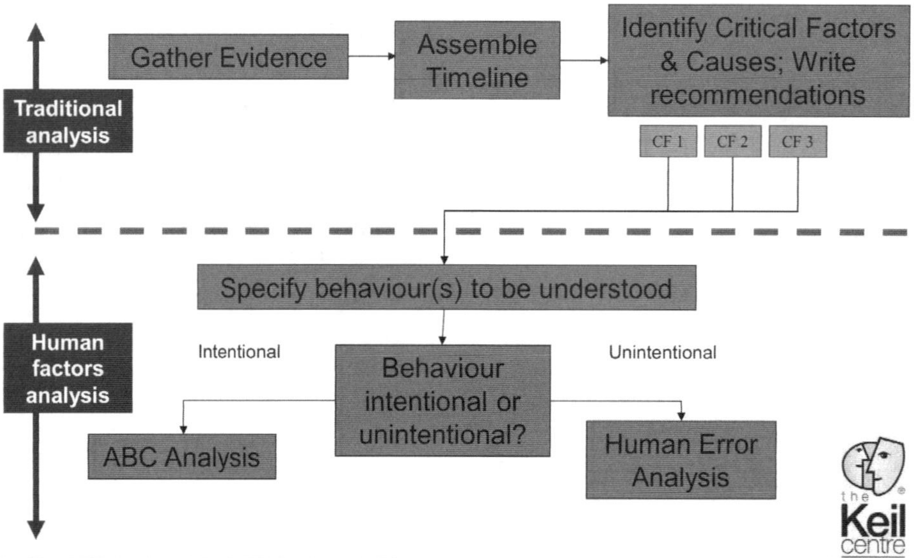

the Keil centre chartered psychologists & ergonomists

The case study..

- Following human factors investigation, focus shifted to onshore managers who did not set up captain for success
    - *Onshore management and commissioning of modifications*
- Communication of lessons learned to all relevant parties
- Question:- "could something ***like*** that happen again?"
- Devised a novel method to answer the question

the Keil centre chartered psychologists & ergonomists

The scenario

- Designed to address question re. "could something ***like*** that happen again?"
- Something the onshore shipping managers might encounter in a typical day
- 12 critical decisions onshore which contributed to incident 6 months before
- Different type of equipment, ship, operation, country…
- Participants asked to record how they ***would*** actually deal with this, and what might influence their decisions
- Worked through individually first time round, then discussed answers as a group
- Explanation of link between scenario, and earlier serious incident
  - Only one person recognised link

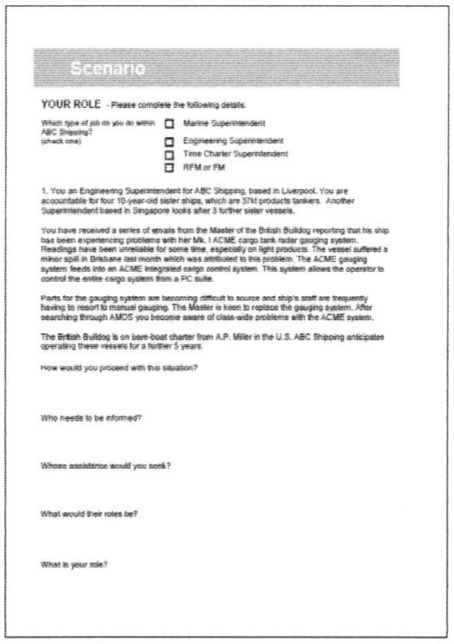

Scenario

YOUR ROLE - Please complete the following details.

Which type of job do you do within ABC Shipping? (check one)

- ☐ Marine Superintendent
- ☐ Engineering Superintendent
- ☐ Time Charter Superintendent
- ☐ RFM or FM

1. You an Engineering Superintendent for ABC Shipping, based in Liverpool. You are accountable for four 10-year-old sister ships, which are 37kt products tankers. Another Superintendent based in Singapore looks after 3 further sister vessels.

You have received a series of emails from the Master of the British Bulldog reporting that his ship has been experiencing problems with her Mk. I ACME cargo tank radar gauging system. Readings have been unreliable for some time, especially on light products. The vessel suffered a minor spill in Brisbane last month which was attributed to this problem. The ACME gauging system feeds into an ACME integrated cargo control system. This system allows the operator to control the entire cargo system from a PC suite.

Parts for the gauging system are becoming difficult to source and ship's staff are frequently having to resort to manual gauging. The Master is keen to replace the gauging system. After searching through AMOS you become aware of class-wide problems with the ACME system.

The British Bulldog is on bare-boat charter from A.P. Miller in the U.S. ABC Shipping anticipates operating these vessels for a further 5 years.

How would you proceed with this situation?

Who needs to be informed?

Whose assistance would you seek?

What would their roles be?

What is your role?

the Keil centre chartered psychologists & ergonomists

Scenario results (23 people)

- Individual Scores (Max = 12)
  - Range: 4-10
  - Average: 6.7
- Critical decision scores
  - Range: 0-22
  - Average: 12.8
- Patterns can be clearly identified

	Critical decision												
Person	1	2	3	4	5	6	7	8	9	10	11	12	Total correct
1													9
2													5
3													10
4													5
5													5
6													7
7													6
8													6
9													8
10													6
11													9
12													4
13													5
14													7
15													9
16													8
17													6
18													6
19													8
20													8
21													5
22													6
23													5
Total correct	18	22	5	0	18	15	7	13	18	10	9	18	

the Keil centre chartered psychologists & ergonomists

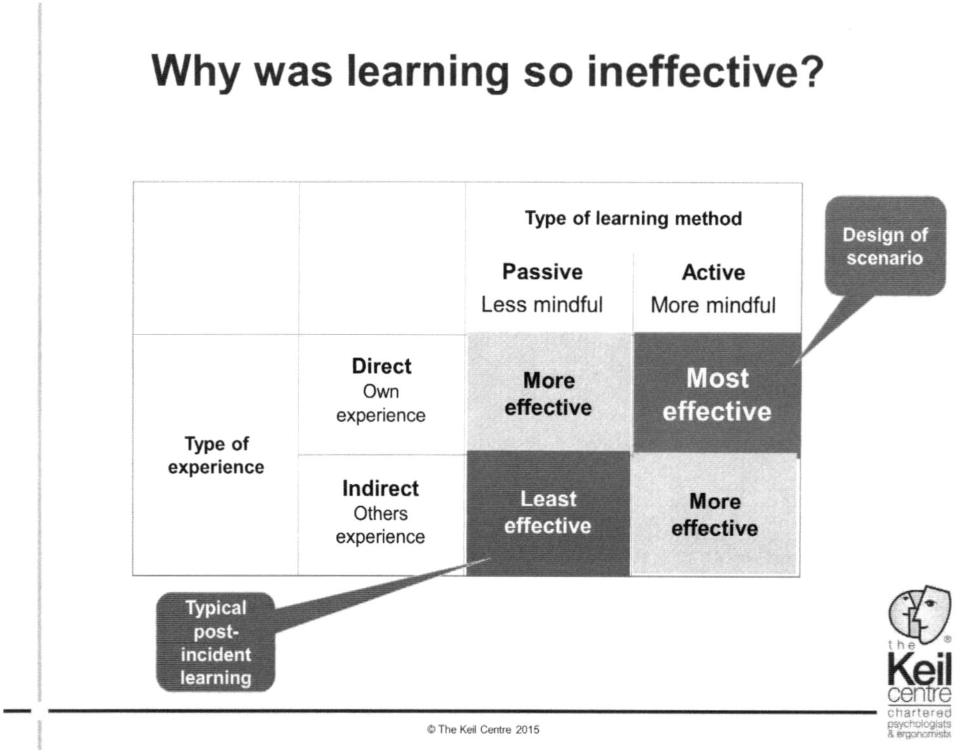

Question	Response	Comments
Views on scenario method	95% positive, liked	"novel and effective"; "better way to discuss incident"; "explore ideas"; "new approach"; "drove the right thinking".
Describe anything new that you learned or realised as a result of completing and discussing the scenario	93% described new learning	Importance of ensuring compliance with critical procedures, and the value of team decision-making & peer review.
Has completing the scenario, and discussing it with your peers, affected how you would approach a similar situation in the future?	61% Yes	Led to more cautious, conservative style of decision-making.
Scenarios better than traditional ways of learning from incidents?	78 % Yes	-

Further application…

- Oil & gas company
- Series of process isolation incidents, over 7 year period
- Large amount of upcoming commissioning, with many isolations & de-isolations
- Constructed scenario based on 8 incidents, with key features of all included
- Initial reactions
  - Engaging, interesting
  - Stimulated debate
  - Much more effective learning method
  - Widespread use supported

the Keil centre chartered psychologists & ergonomists

17

Implications

		Type of learning method	
		Passive Less mindful	**Active** More mindful
Type of experience	**Direct** Own experience	**More effective**	**Most effective**
	Indirect Others experience	**Least effective**	**More effective**

Use more active and direct learning methods

Typical post-incident learning

the Keil centre chartered psychologists & ergonomists

19

Summary

- A more sophisticated understanding of scope of human factors, and the many & varied influences on behaviour is required
- Development of internal company expertise is slowly taking place
- Integration of human factors into design, operations, maintenance and decommissioning is necessary

the Keil centre chartered psychologists & ergonomists

20

IChemE Human Factors professional development course

- Joint initiative with The Keil Centre
- 4 x 2-day modules, spread over 12 months
- Covers main "top-ten" HF topics
- Five European courses now finished, three in Australia
  - 150 delegates in Australia
- Melbourne course commences in June 2015

the Keil centre chartered psychologists & ergonomists

Further information & resources

21

- "Reducing Error and Influencing Behaviour"
  - Download free at www.hsebooks.co.uk - search for HSG48
- UK Health and Safety Executive HOF web-site
  - http://www.hse.gov.uk/humanfactors/index.htm
- US Chemical Safety Board Safety Videos
  - http://www.csb.gov/videoroom/videorequest.aspx?
- Energy Institute HOF briefing notes
  - http://www.energyinst.org.uk/index.cfm?PageID=811
- Step-Change in Safety
  - "Human Factors – how to take the first steps" – search in Google

the Keil centre chartered psychologists & ergonomists

Thinking: A user's guide

Martin Anderson C.ErgHF FIEHF
Manager Human Factors
Woodside

AIChE CCPS - Asia Pacific Conference
Perth
10 February 2015

Disclaimer and important notice

This presentation contains forward looking statements that are subject to risk factors associated with oil and gas businesses. It is believed that the expectations reflected in these statements are reasonable but they may be affected by a variety of variables and changes in underlying assumptions which could cause actual results or trends to differ materially, including but not limited to: price fluctuations, actual demand, currency fluctuations, drilling and production results, reserve estimates, loss of market, industry competition, environmental risks, physical risks, legislative, fiscal and regulatory developments, economic and financial market conditions in various countries and regions, political risks, project delay or advancement, approvals and cost estimates.

All references to dollars, cents or $ in this presentation are to US currency, unless otherwise stated.

References to "Woodside" may be references to Woodside Petroleum Ltd. or its applicable subsidiaries.

10 Feb 2015 Thinking: A user's guide

Which centre circle is bigger?

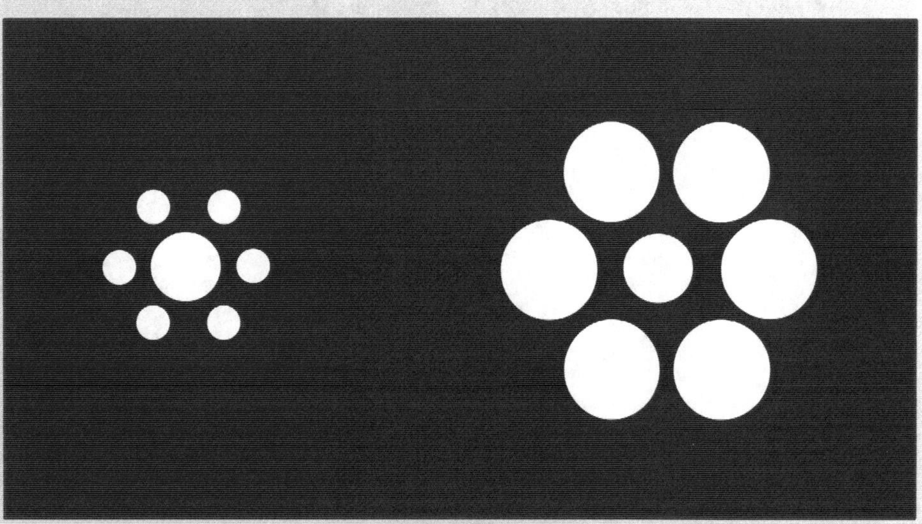

The mind as a powerful tool

Aoccdrnig to rsceearch at Nttoingahm Uinervtisy, it deosn't mttaer in waht oredr the ltteers in a wrod are, the olny iprmoetnt tihng is taht the frist and lsat ltteer be at the rghit pclae. The rset can be a total mses and you can sitll raed it wouthit a porbelm. Tihs is bcuseae the huamn mnid deos not raed ervey lteter by istlef, but the wrod as a wlohe.

Dr Rawlinson, G. E. (1976), University of Nottingham, UK.

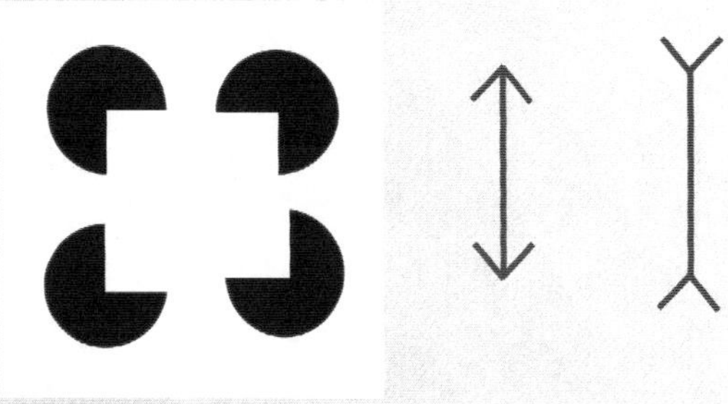

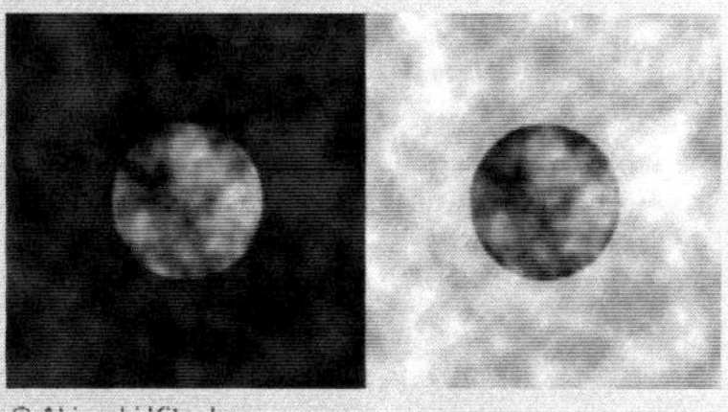

© Akiyoshi Kitaoka

© Akiyoshi Kitaoka

The "Top 10" human factors topics

Optimising human performance	Training & competence
Design of equipment, processes, tasks & environment	Procedures
Organisational change	Safety critical communications (including handovers, permits & alarms)
Supervision	Fatigue & shiftwork
Staffing levels & workload	Organisational & safety culture

"Human factors uses what we know about people, organisations and work design to influence performance"

Thinking: the psychology

- What we "see" or "hear" is filtered
- We can't hold everything in our working memories
- There are bottlenecks at certain points, just like in an industrial process
- We rely on 'mental models'
- We delegate routine tasks to our autopilot
- We have thinking habits and 'shortcuts'

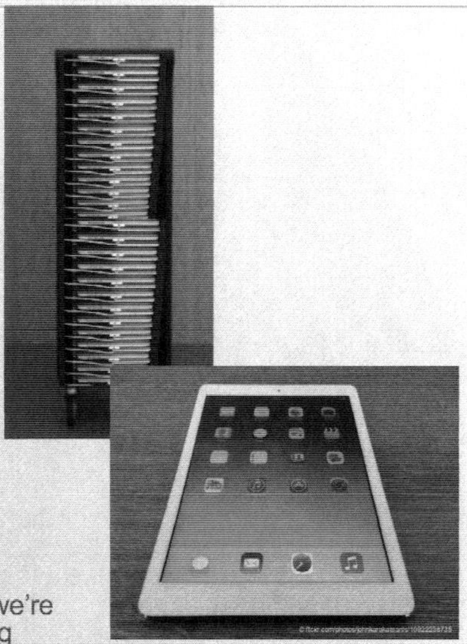

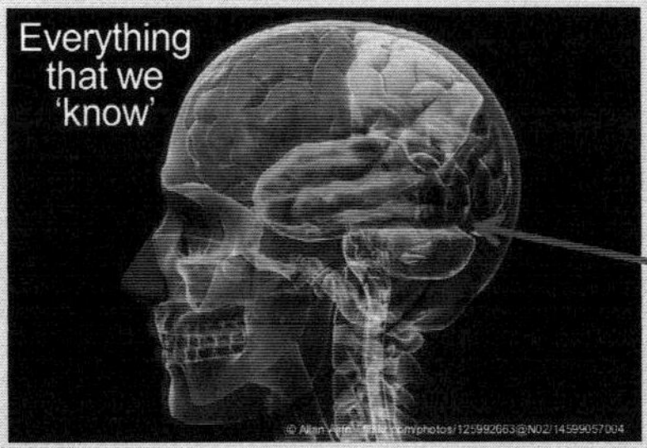

What we're thinking about right now

WE KEEP EACH OTHER SAFE

Fast and slow thinking

Fast thinking

- our default mode
- automatic
- without questioning
- prone to assumptions and and rules of thumb

S-l-o-w thinking

- deliberate and rational
- takes effort
- has to be switched on

WE KEEP EACH OTHER SAFE

"Mind traps": Thinking errors

- Mind traps are common
- Mainly occur during fast thinking
- They are coping mechanisms:
  - helping our minds make sense of huge amounts of information
  - conserving the brain's energy

© David Goehring - flickr.com/photos/carbonnyc/2614226455

Mind traps in our everyday language:

- "wishful thinking"
- "jumping to conclusions"
- "you see what you want to see"
- "not thinking it through"
- "your mind playing tricks on you"

WE KEEP EACH OTHER SAFE

A common mind trap: Confirmation bias

- We prefer information that matches our expectations
- We seek out information that confirms our views, or a decision that we have made
- We pay attention to what we expect to see
- We may avoid or discount information that contradicts
- Problematic or ambiguous information is rationalised away

- These tendencies become stronger when we're under pressure

Confirmation bias in personal decisions

- consider this in your decisions, such as major purchases (e.g. house, car, TV, or investments)

Shubert Ciencia © flickr.com/photos/bigberto/2770838880

WE KEEP EACH OTHER SAFE

Macondo, 2010: Example of confirmation bias

- Well integrity test:
  - To check that the cement seal was working prior to moving the rig
  - Very unlikely that the well would fail the test
  - Test seen as a box-ticking exercise, as the cement engineers declared it a success
  - The team were looking for <u>confirmation</u> that the well was sealed, not investigating whether it was or not
  - Indications were that the well failed the integrity test: oil & gas were passing the seal
  - Explained away the unexpected test results with 'bladder effect'
  - Applied another technique to find the confirmation that they were seeking
  - Once declared a success, team put behind them any concerns about the test

"Once these processes are taken into account, the faulty decisions made by the Macondo group become entirely understandable, terrifyingly so" (Hopkins, 2012)

WE KEEP EACH OTHER SAFE

10 Feb 2015 Thinking: A user's guide 11

Numerous mind traps: Flight 447 (1 June 2009)

- Pilots warned cabin crew about possible turbulence
- Icing affected airspeed indicators
  - Autopilot disengaged
  - Auto-thrust disengaged

Photo credit: © flickr.com/photos/airlines470/14566938386

Pilots flying manually

- Aircraft rolled right due to turbulence
- Pilot over-corrected: rolled left and right
- Pilot moved the nose-up (to 40%)
- Thrust at maximum
- Ascent & reduced speed created aerodynamic stall
- "Stall" warning sounded 75 times
- Descended steeply (11,000 ft per minute)
- Remained stalled until crash (3.5 minutes)
- **Note: engines responded normally throughout**

WE KEEP EACH OTHER SAFE

10 Feb 2015 Thinking: A user's guide

"*Total loss of cognitive control*" leads to a loss of physical control . . .

Cockpit Voice Recorder:

- *"We still have the engines! What the hell is happening? I don't understand what's happening*
- *Damn it, I don't have control of the plane, I don't have control of the plane at all!*
- *What the hell are you doing?*
- *We've totally lost control of the plane. We don't understand at all... We've tried everything*
- *What do you think? What do you think? What should we do?*
- *Well, I don't know!"*

WE KEEP EACH OTHER SAFE

Air France flight 447: Mind traps

- **Attentional tunnelling**: focussed on altitude and electronic alerts
- **Expectation bias**: pilots anticipated turbulence - and associated the shaking and loss of altitude with the weather (rather than with stalling)
- **Reliance on technology**: thought flight computers would prevent a stall
- **Selective perception**: ignored the clear and repeated "stall" warning
- **Automatic response**: trained response to unusual situations was to apply full nose-up (not applicable to this situation)
- **Confirmation bias**: discounted conflicting cues, held onto the notion that they had lost all instruments (rather than consider a stall)
- **Overconfidence/optimism effect**: situations usually turn out well
- **Team dynamics**: left seat not working well with the right seat

Normalisation of risk (Airbus and Air France):

- many similar previous events were not acted upon

WE KEEP EACH OTHER SAFE

Managing mind traps

- What can we do right now?
- What can you do differently during a complex task, unusual situation or incident?

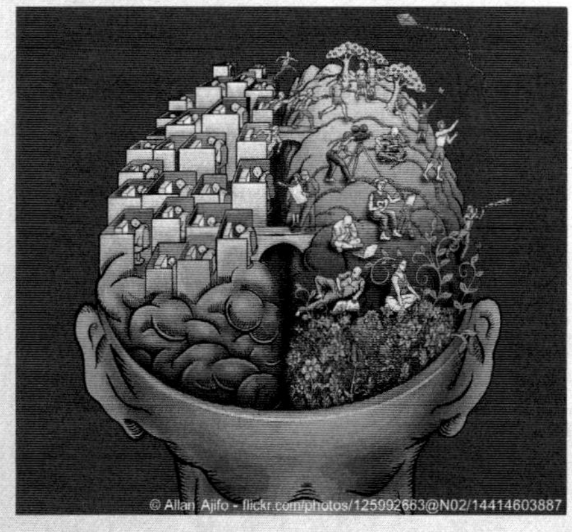

Managing mind traps: What can we do right now?

Remove influences that encourage *fast thinking*:

- distractions
- excessive workload
- stress
- fatigue
- time pressures
- etc.

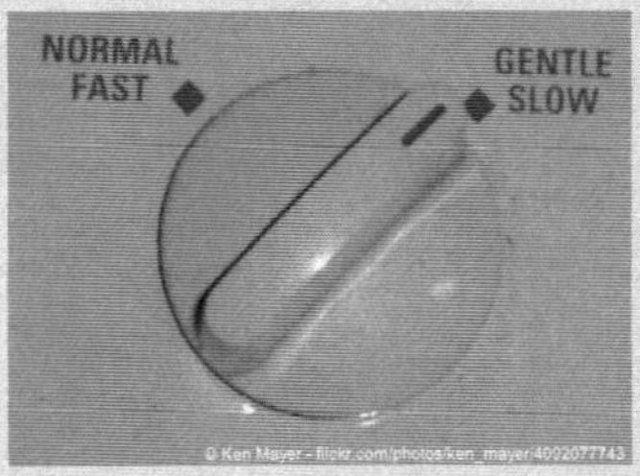

Inoculate the organisation:

- Improve awareness of potential mind traps
- Review previous decisions: are you susceptible to certain mind traps?
- Reduce complexity
- Training in upsets and emergencies
- Non-technical skills training, such as team work and decision-making

Managing mind traps: During tasks and 'situations'

Aggressively seek out and understand information that disagrees with your existing beliefs/decisions

During a complex task, unusual situation or incident, try to **disengage your autopilot** and **s-l-o-w your thinking:**

- Step back and reflect: "Take 5"
- Re-consider decisions as new data becomes available
- Create two options and compare the pros and cons
- What would your 2nd and 3rd choices be?
- Ask open questions ("why, how, what"?)
- Listen to strong opposing views
- Make your assumptions explicit - write them down or discuss them with others

- Mind traps are often subconscious - **so look out for them in others and intervene**

woodside

Step back 5x5

Take 5 minutes to make the job safe

Before you start work, step back and:

1. **Think**
   Think through the task and surroundings
2. **Identify**
   Identify any hazards
3. **Assess**
   Assess the risks and select the controls and recovery measures
4. **Change**
   Make the changes
5. **Review**
   Review steps 1 - 4

When it is safe, proceed with the job.

If you feel that the job is unsafe, stop and discuss with your supervisor.

KEEP CALM AND ASSUME NOTHING

Theme	Everyone	Supervisors	Managers/Executives
Standards	Follow rules	Ensure compliance	Set high standards
Communication	Speak up	Encourage the team	Communicate openly
Risk management	Be mindful	Promote risk awareness	Confront risk
Involvement	Get involved	Involve the team	Involve the workforce

WE KEEP EACH OTHER SAFE

10 Feb 2015 Thinking: A user's guide 17

ConocoPhillips Australia

Workforce Engagement in Human Factors and Process Safety

Mark Leigh
Team Leader, Integrity and Process Safety

Cautionary Statement

The following presentation includes forward-looking statements. These statements relate to future events, such as anticipated revenues, earnings, business strategies, competitive position or other aspects of our operations or operating results. Actual outcomes and results may differ materially from what is expressed or forecast in such forward-looking statements. These statements are not guarantees of future performance and involve certain risks, uncertainties and assumptions that are difficult to predict such as oil and gas prices; operational hazards and drilling risks; potential failure to achieve, and potential delays in achieving expected reserves or production levels from existing and future oil and gas development projects; unsuccessful exploratory activities; unexpected cost increases or technical difficulties in constructing, maintaining or modifying company facilities; international monetary conditions and exchange controls; potential liability for remedial actions under existing or future environmental regulations or from pending or future litigation; limited access to capital or significantly higher cost of capital related to illiquidity or uncertainty in the domestic or international financial markets; general domestic and international economic and political conditions, as well as changes in tax, environmental and other laws applicable to ConocoPhillips' business and other economic, business, competitive and/or regulatory factors affecting ConocoPhillips' business generally as set forth in ConocoPhillips' filings with the Securities and Exchange Commission (SEC).

Use of non-GAAP financial information – This presentation may include non-GAAP financial measures, which help facilitate comparison of company operating performance across periods and with peer companies. Any non-GAAP measures included herein will be accompanied by a reconciliation to the nearest corresponding GAAP measure in an appendix.

Cautionary Note to U.S. Investors – The SEC permits oil and gas companies, in their filings with the SEC, to disclose only proved, probable and possible reserves. We use the term "resource" in this presentation that the SEC's guidelines prohibit us from including in filings with the SEC. U.S. investors are urged to consider closely the oil and gas disclosures in our Form 10-K and other reports and filings with the SEC. Copies are available from the SEC and from the ConocoPhillips website.

Workforce engagement in Process Safety

- What is Process Safety?
- Managing process safety barriers
- How to drive an improvement in Process Safety awareness
- Communication Tools
- Metrics and results

What is Process Safety?

- Management and control of process hazards in a facility that have the potential to adversely impact on people, property or the environment.
- Three key areas of Process Safety - design & engineering, asset integrity and operating integrity.

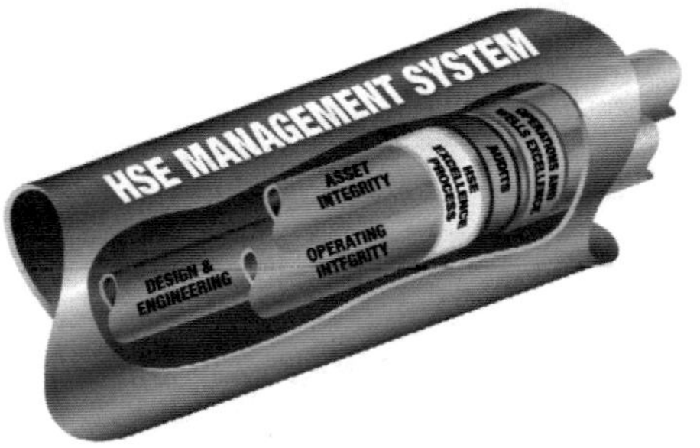

Major accident events are rare – right?

➨ Major accident events are not as rare as you might think.

Incidents 1900-2013

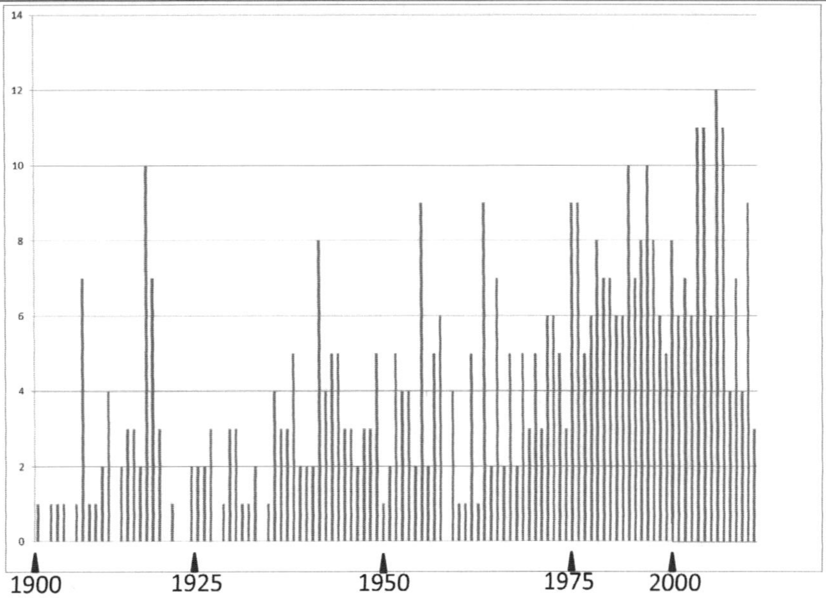

Average >4 MAE per year
8/year since 2000

Incidents 1900-2013 and fatalities

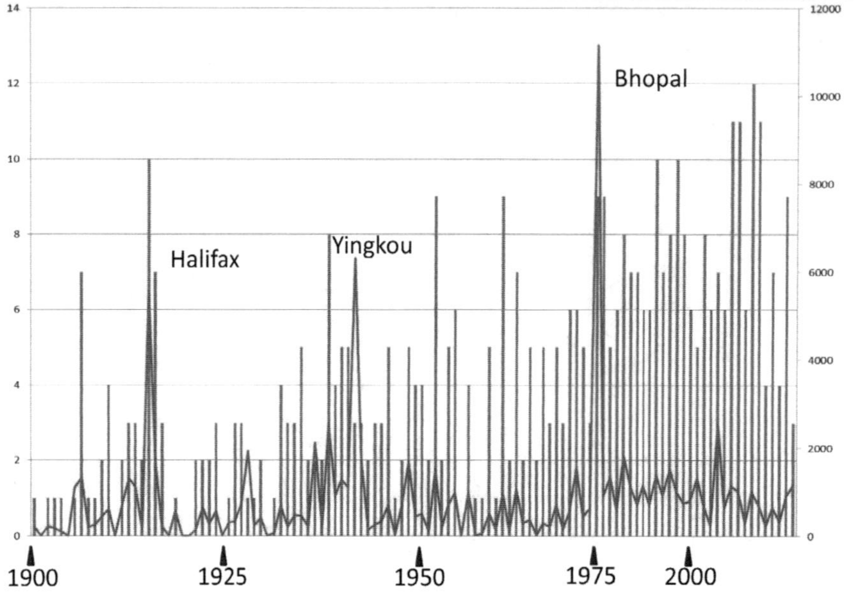

Average >4 MAE per year
8/year since 2000

Average 850 fatalities/year
11,500 since 2000 ~ 880/year

Workforce engagement

- Executive leadership
  - Policy
  - Governance
  - Audit
- Management
  - Goals
  - Strategies
  - Review
  - KPIs

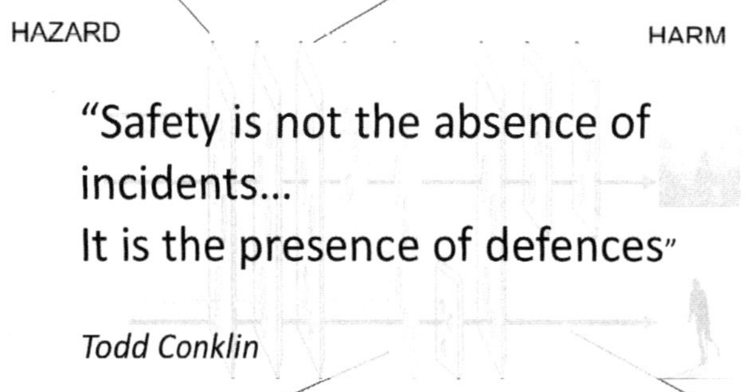

"Safety is not the absence of incidents...
It is the presence of defences"

Todd Conklin

- Supervisors
  - Compliance
  - Field leadership
  - Permits
- Workforce
  - Welders
  - Engineers
  - Purchasing
  - Technicians

Which Process Safety barrier Do YOU own??

Source: Wikimedia Commons

Process Safety 5-year Plan

	2014	2015	2016	2017	2018
Management systems	MOC program improvements				
	Risk database consolidation				
	Human Factors – Investigation	Data Analysis	Action Implementation		
	Safety Case resubmission				
Physical barriers	Asset Integrity projects				
	PHA program				
	PHA Action items implementation				
	SCE performance verification				
	Temporary deviation program improvements				
Human barriers	PS, hazard and risk awareness improvements				
	Improved information management systems				
	Alarm management				
	TA engagement and development				
	OI and human factors - operations competence and procedures				
	Improve org change management				

Five Year Plan detailed tool

From Overview plan

Metrics

Future years actions

Group	OE Reference	Topic	Sub-Topic	2014	Owner	Status	2015	
		Actions this year 47		87%	in-hand	41		
Management Barriers	1.408; 3.401; 4.108; 4.300; 5.217	Improve MOC program	MOC	Review MOC/Risk Meeting agendas	Cones	1	Develop CI Project to move TODs to SAP	Transfer TC
Management Barriers	2.212; 3.105	Improve MOC program	MOC	Harmonise DLNG and Bayu-Undan meetings	Cones	1		
Management Barriers	5.208	Improve MOC program	MOC	Improve use of risk tab in ZA notification	Jennings	1		
Management Barriers	5.207	Improve MOC program	MOC	Identify ownership of PSSR program	Cones	1		
Management Barriers	5.105; 3.203	Risk Consolidation and interdependence	Risk	Link together various risk tools into common display/report	Leigh	1	Identify integrated risk platform Develop CI project Continue quarterly senior management briefings	Implement inte Continue quart
Management Barriers	4.112; 5.111	Risk Consolidation and interdependence	Risk	Continue quarterly senior management briefings	Leigh	1		
Management Barriers	4.103; 4.217; 6.309	Human Factors	Human Factors	Evaluate improved investigation program to identify HF factors	Duk	1	Implement HFAT Analyse investigation data Identify common HF issues Deliver HF training modules	Develop HF imp
Management Barriers	4.201; 4.213; 4.214; 2.213	Human Factors	Human Factors	Develop standard HF briefing pack for supervisors - do we need to wait for COP global messaging	Leigh	0		
Management Barriers	5.218; 5.219	Bayu Undan Safety Case	Analysis	Update Bayu Undan Formal Safety Assessments to account for changes to the facility since last Safety Case update.	Cross	1	Periodic information and awareness sessions on Safety Case Continued SIMOPS/COMOPS audits	Continued SIM External review c
Management Barriers	3.203; 4.109	Bayu Undan Safety Case	Analysis	Update Bayu Undan Quantitative Risk Assessment to account for changes in field as well as updates to leak and ignition probability historical data sets	Cross	1		

Detailed action and owner

References relevant Ops Excellence section

Example communications in use in Australia-West

- HSE Theme of the week
- Flat-screen messages
- Newsletters
- Boardwalks
- Safety moments
- Shift presentations
- Websites and Wiki pages
- WAVES observations (BBS)
- Corporate Life Saving Rules
- Town-halls
- MOC open day
- Morning tea
- Competitions
- HAZOPs/PHA
- Bi-weekly risk review meetings
- Metrics

- Adjust existing systems to address PS

Build on your existing OHS mindset

- Personnel Safety
  - Line of fire
  - PPE
  - Ergonomics
  - Chain integrity
  - Housekeeping

- Process Safety
  - Correct torquing on bolts
  - Compatible bolt & flange material
  - Correct gasket
  - Flange face condition
  - Correct lubricants

- Build on existing strengths in OHS

Communication tool #2 - Newsletter

- Use human factors thinking
- Not this

Shutdown Preparation

Lorem ipsum dolor sit amet, consectetur adipiscing elit. Integer nec odio. Praesent libero. Sed cursus ante dapibus diam. Sed nisi. Nulla quis sem at nibh elementum imperdiet. Duis sagittis ipsum. Praesent mauris. Fusce nec tellus sed augue semper porta. Mauris massa. Vestibulum lacinia arcu eget nulla. Class aptent taciti sociosqu ad litora torquent per conubia nostra, per inceptos himenaeos. Curabitur sodales ligula in libero. Sed dignissim lacinia nunc. Curabitur tortor. Pellentesque nibh. Aenean quam. In scelerisque sem at dolor. Maecenas mattis. Sed convallis tristique sem. Proin ut ligula vel nunc egestas porttitor. Morbi lectus risus, iaculis vel, suscipit quis, luctus non, massa. Fusce ac turpis quis ligula lacinia aliquet. Mauris ipsum. Nulla metus metus, ullamcorper vel, tincidunt sed, euismod in, nibh. Quisque volutpat condimentum velit.

Focus Topic - Lifting Plans

Class aptent taciti sociosqu ad litora torquent per conubia nostra, per inceptos himenaeos. Nam nec ante. Sed lacinia, urna non tincidunt mattis, tortor neque adipiscing diam, a cursus ipsum ante quis turpis. Nulla facilisi. Ut fringilla. Suspendisse potenti. Nunc feugiat mi a tellus consequat imperdiet. Vestibulum sapien. Proin quam. Etiam ultrices. Suspendisse in justo eu magna luctus suscipit.

Sed lectus. Integer euismod lacus luctus magna. Quisque cursus, metus vitae pharetra auctor, sem massa mattis sem.

- But this!

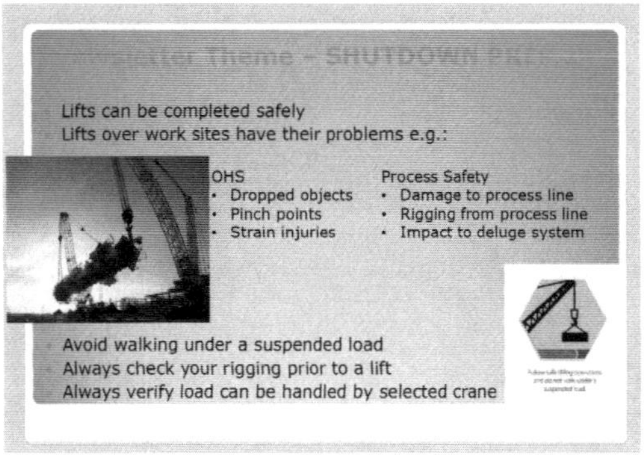

Metrics

- Stay focused on the base of the pyramid

Metrics 2

- Choose measures appropriate for your business

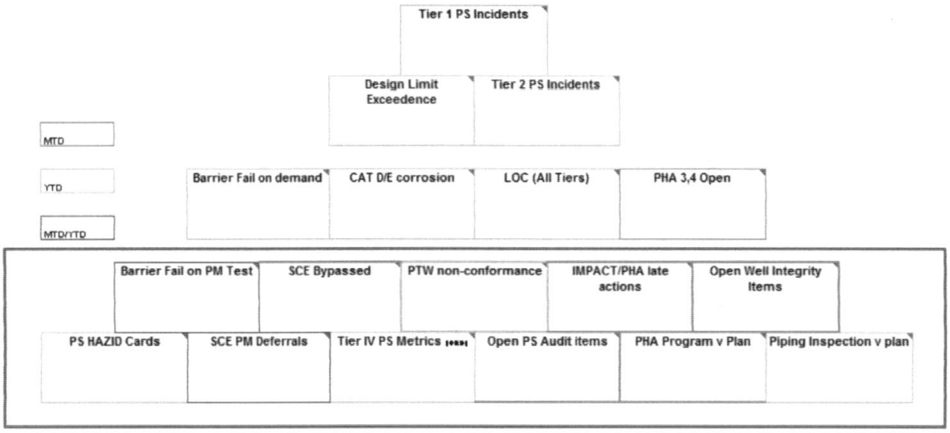

Conclusions and key messages

Which Process
Safety barrier
Do YOU own??

Source: Wikimedia Commons

Alarm Management
within Major Hazard Facilities

Albert Romano

Outline

- Background to our business
- Importance of alarm management
- Are there too many alarms?
- How to manage alarms
- Case study

proud history – great future

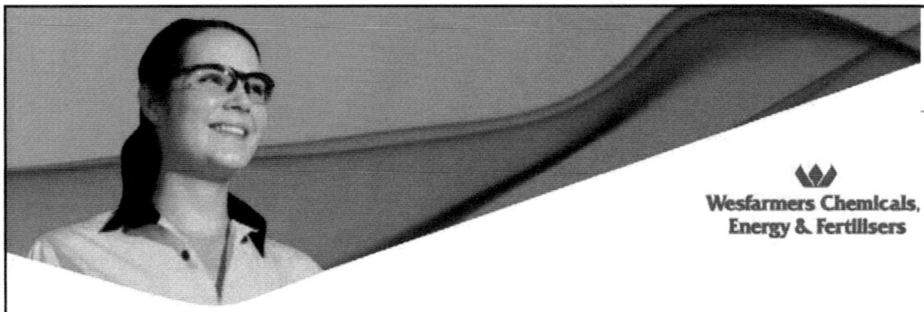

Wesfarmers Chemicals, Energy & Fertilisers

WesCEF DIVISION PROFILE

Wesfarmers Chemical, Energy and Fertilisers (WesCEF) operates a diverse range of **chemical, gas** and **fertiliser** businesses that service a range of sectors in both domestic and international markets.

proud history – great future

CSBP Chemicals COMPANY PROFILE

Our businesses manufacture and supply:

Ammonia, ammonium nitrate and industrial chemicals to the WA mining sector

Ammonium nitrate to the QLD coal mining industry though QNP, our 50% JV with Dyno Nobel Asia Pacific

Sodium cyanide to the WA and international gold mining sector through Australian Gold Reagents (AGR), our 75% joint venture with Coogee Chemicals.

Polyvinyl chloride resin (PVC) and wood-plastic composites to the Australian industrial sector through Australian Vinyls and its subsidiary ModWood.

proud history – great future

CSBP Fertilisers COMPANY PROFILE

CSBP Fertilisers manufactures, imports and distributes inorganic phosphate, nitrogen and potassium based fertilisers in blended, compound and liquid form for the WA agricultural sector.

At its Kwinana operation, CSBP operates **manufacturing plants** for the production of **compound granular fertiliser**, **superphosphate granular fertiliser**, and **liquid fertiliser**.

proud history – great future

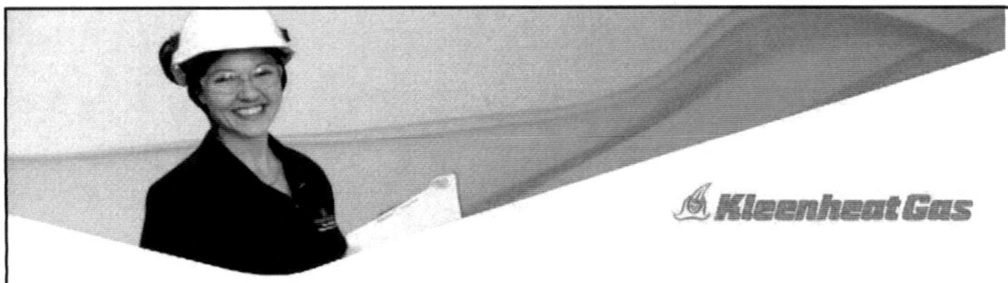

Kleenheat Gas COMPANY PROFILE

Kleenheat Gas extract, distributes and markets LPG and is a leading provider of gas services to residential, commercial, industrial, rural, and automotive markets across Australia.

Kleenheat operates **LPG** and **LNG manufacturing plants** at Kwinana.

The Kleenheat Gas sells into:

Natural gas retailer to homes and businesses.

LPG gas supply to commercial areas.

Kwik-Gas – cylinder exchange programme.

AutoGas – automotive fuel source.

EVOL LNG – fuel source for heavy vehicles, industrial and remote power generation.

What is alarm management and why is it important ?

proud history – great future

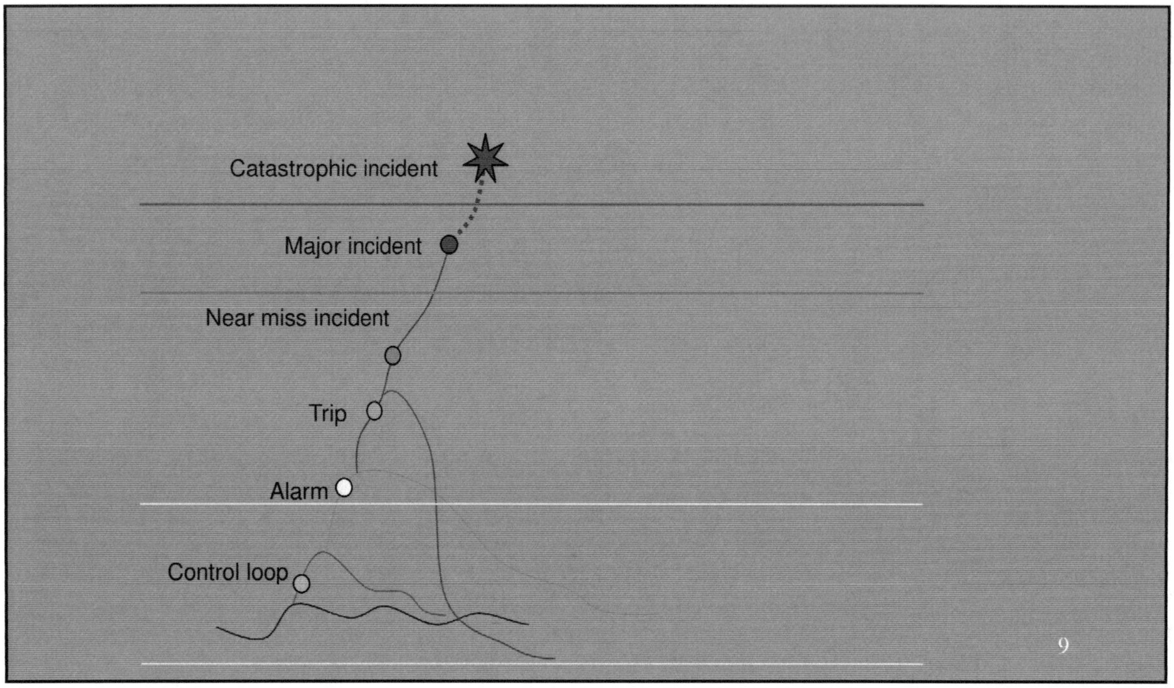

Why manage alarms?

Human error *is a **DIRECT** cause to 42 per cent of all incidents* [1]

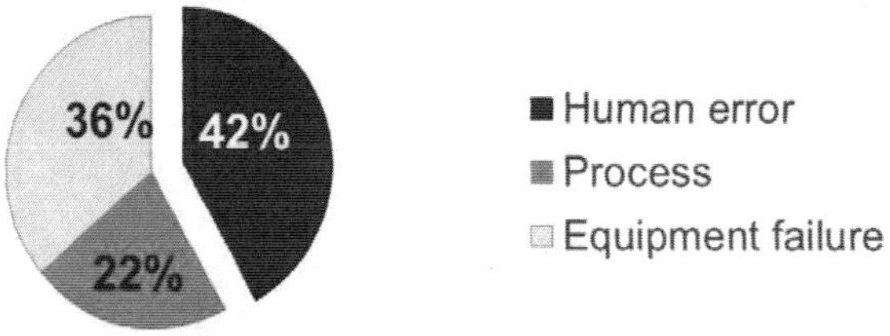

Experience has shown human error is a **CONTRIBUTING** cause to more than 95 per cent of all incidents [2]

[1] U.K. Health and Safety Executive, [2] DuPont Australia

proud history – great future

Why too many alarms?

What is an alarm (supposed to be)?

Key words that define an alarm are; an operator response is required

Definition of an alarm[1]:

- An audible and/or visual means of indicating to the **operator**
- an equipment malfunction, process deviation or any other **abnormal condition**,
- **requiring a response**

		Operator action	
		action required	no action required
Plant condition	abnormal		
	expected		
	normal		

[1] ANSI/ISA-18.2 2009 Management of Alarm Systems for the Process Industries (ISA-18.2)

Rationalisation process

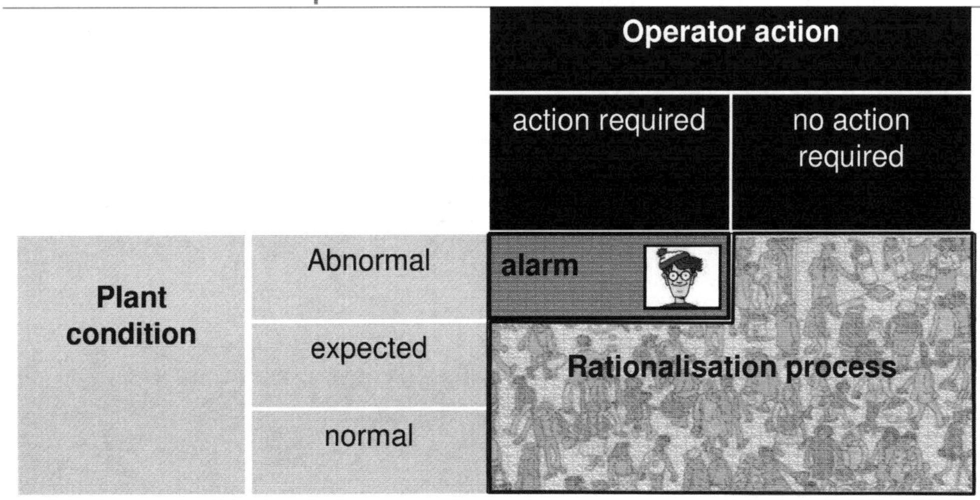

proud history – great future

Rationalisation process

Rationalisation involves a process that focuses on identifying necessary / unnecessary alarms

(i) **Fix** unnecessary alarms:

- Validate/remove
- document
- prioritise
- classify
- set limits

(ii) **Suppress** unnecessary alarms

Can do this manually or use software to assist, but a review is required

proud history – great future

How to alarm manage

Success requires following an improvement process.

Basic outline:

1. Develop, adopt, and maintain an alarm philosophy
2. Collect data and benchmark alarm performance
3. Perform "bad actor" alarm resolution
4. Perform documentation and rationalization
5. Implement alarm auditing
6. Implement applicable elements of real-time alarm management
7. Apply controls to maintain improvements

proud history – great future

Case Study - Alarm management activities

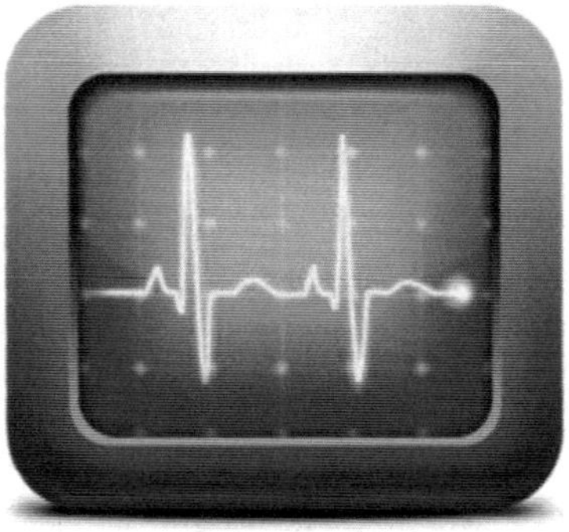

proud history – great future

Questions
www.csbp.com.au
www.wescef.com.au
www.wesfarmers.com.au
albert.romano@csbp.com.au

PTTEP

Striking the Balance Between Personal Safety and Process Safety

Robin Wright, Safety, Security, Health and Environment Manager

1

PTTEP

Outline

- Introduction
- MAE prevention: the "Line of Sight" tool
- Safety Critical Element Performance Standards
- The "Case to Operate"
- Safety Metrics
- A culture of "mindfulness"
- Summary

2

Lions and Tigers

3

Taming the Two Beasts

- The Lion - Personal Safety
  - Hunts in packs
  - Is often visible (audible etc)
  - Common in its own environment
  - Is a killer
  - Can never be fully tamed
  - Can never be left unguarded
  - Needs constant vigilance

Turn your back on either at your peril

- The Tiger – Process Safety
  - Hunts alone
  - Is secretive and hard to detect (well camouflaged)
  - Not commonly seen
  - Is a killer
  - Can never be fully tamed
  - Can never be left unguarded
  - Needs constant vigilance

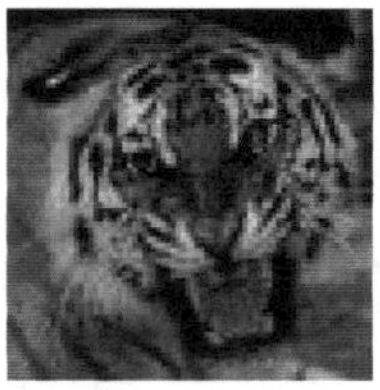

Montara

The Montara Action Plan

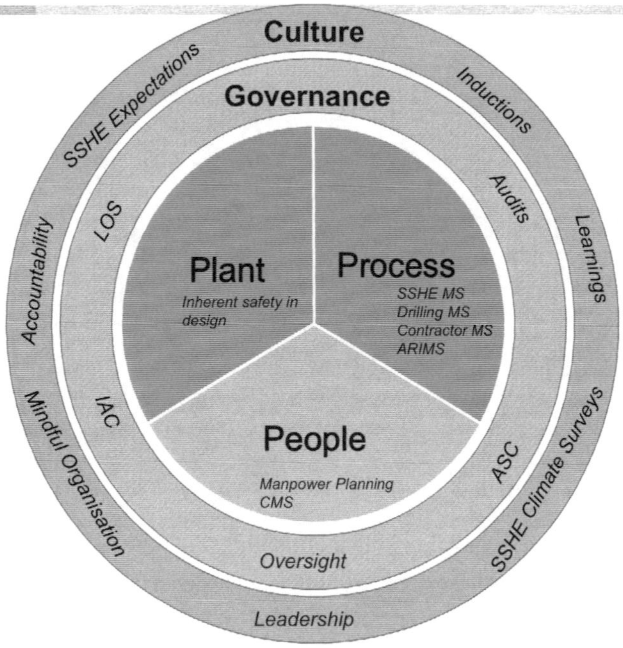

6

Risk Management

Hazard Identification

JHA — Risk Assessment

Controls

Bow-ties

Safety Critical Elements

Performance Standards

Life Saving Rules

Line of Sight – The Concept

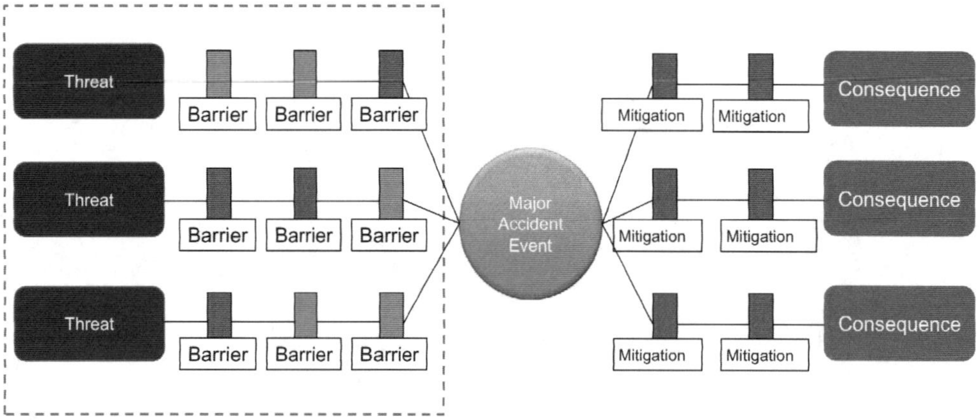

FOCUS OF LINE OF SIGHT SYSTEM

8

PTTEP

Line of Sight - Development

SELECTED MAEs

- Unignited blowout
- Loss of stability
- Large scale fire
- Toxic gas release
- Helicopter Operations

SELECTED BARRIERS

- Equipment meets HA
- Fire control equipment functional
- **Fire precautions for engine room**
- Functional PA
- Fire fighting equipment deviation approved
- Effective hot PTW
- Lifeboats operational
- Competent fire team

SELECTED CHECKS

- Explosives bunkers in designated areas
- Fire dampers functional
- Fuel control system tested and serviced
- Remote engine shut down system tested

9

PTTEP

Line of Sight – In Operation

MAEs | Barriers | Checks

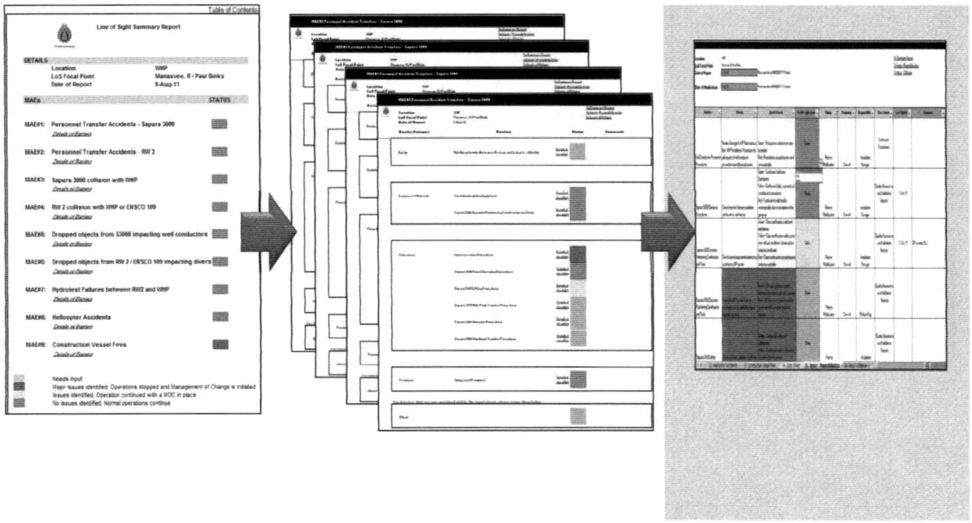

10

Line of Sight – has been effective

ILLUSTRATIVE

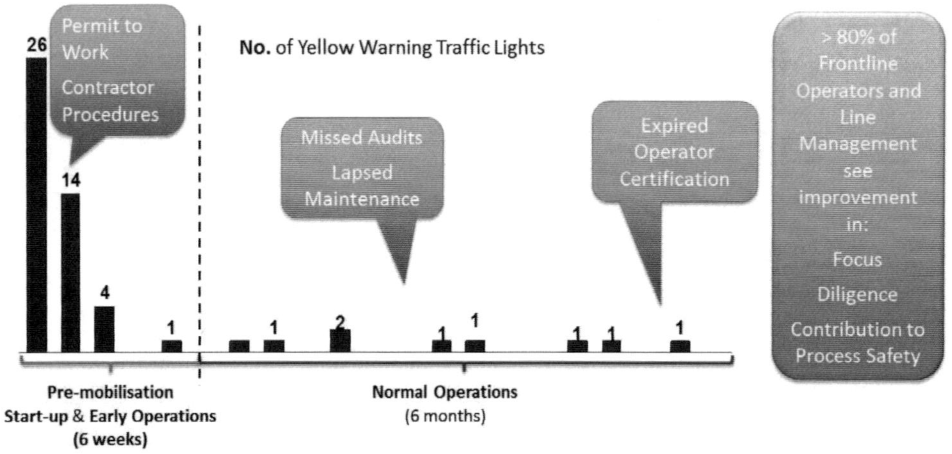

11

Case to Operate

Key features of the Case to Operate

- Issued within 12 hours
- Registered in offshore database
- Risk assessment conducted with the team
- If risk can be mitigated, approved by management
- Continued operation within the agreed period
- Problem rectified in the agreed period
- Validity 1 month; extension < 6 months

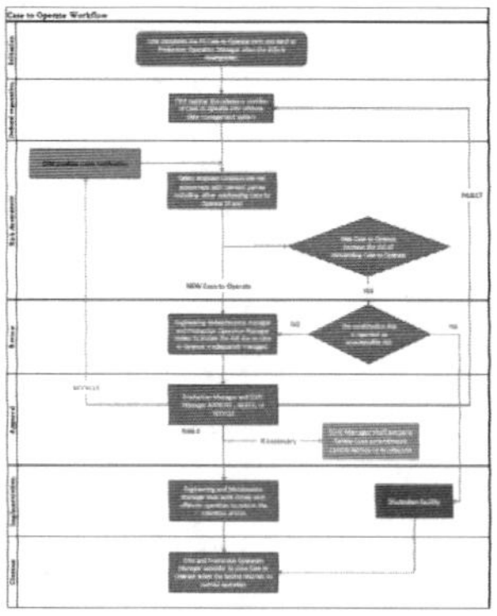

12

Safety Critical Elements

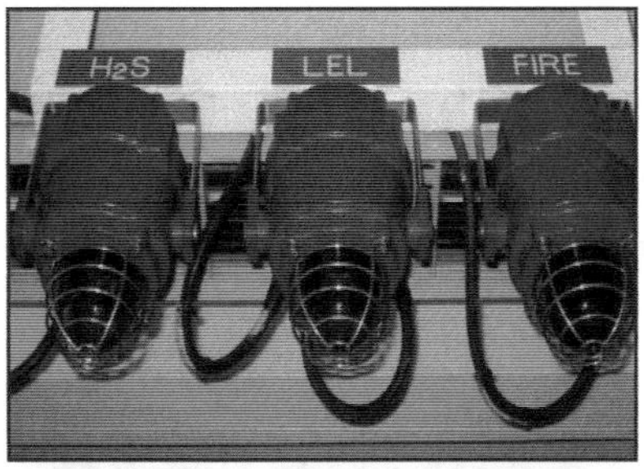

Safety Critical Elements:

- Failure could cause or substantially contribute to a major accident event.
- Purpose is to prevent or limit the effect of a major accident.

Examples:

- Fire detection
- Inert gas system
- Emergency power
- Temporary refuge
- Egress routes

Performance Standards

- **Acceptance Criteria** - practical performance criteria for each critical element/sub-element.
- **Assurance (and Validation) Tasks** - maintenance activities to help demonstrate achievement of acceptance criteria.

- **A PS establishes**:
  - Goal
  - Scope
  - MAEs
  - Functionality
  - Reliability
  - Survivability
  - Availability

MONTARA PROJECT OPERATIONS PERFORMANCE STANDARD

Title	Emergency Shutdown						
Code	MON-WHP-F71	**Revision**	B	**Revision Date**	26/09/2011	**Custodian**	

GOAL:	**To accomplish safe shutdown and isolation of plant and equipment to prevent or mitigate the effects of a major accident event.**
SCOPE:	• All Shutdown Valves (SDVs) as shown on the P&IDs [1]. • Manual Shutdown Initiators • Integrated Control and Safety System (ICSS) including. ○ WHP Emergency Shutdown Systems ▪ WHP Emergency Shutdown System Logic Solver (ESDLS) and associated I/O hardware ▪ Safety Instrumented Logic Solvers and associated I/O hardware ▪ Other IPF assessed Logic Solvers and associated I/O hardware ○ Loop devices including sensors and outputs to final elements • Fire and Gas shutdown initiator inputs to the ESD system as identified on the Field-wide integrated Cause & Effects [1]. • All Electrical isolations controlled by the ESD system. • High Integrity Pressure Protection System (HIPPS). *Note. ESD Valves included as part of the scope are fail safe hence associated hydraulics/pneumatics for the ESD valves are not included in scope.* **Exclusions:** • Reservoir isolation valves including wellhead valves are covered under "Subsea Well Reservoir Isolation" (MON-FPSO-F22) and "Surface Well Reservoir Isolation" (MON-WHP-F72) • Relief Valves covered under "Relief System" (MON-WHP-F75)
MAEs:	• All MAEs that could potentially require shutdown of the WHP: MAE-2 - MAE-12, MAE-14 - MAE-16, MAE-18 - MAE-19

Safety Metrics

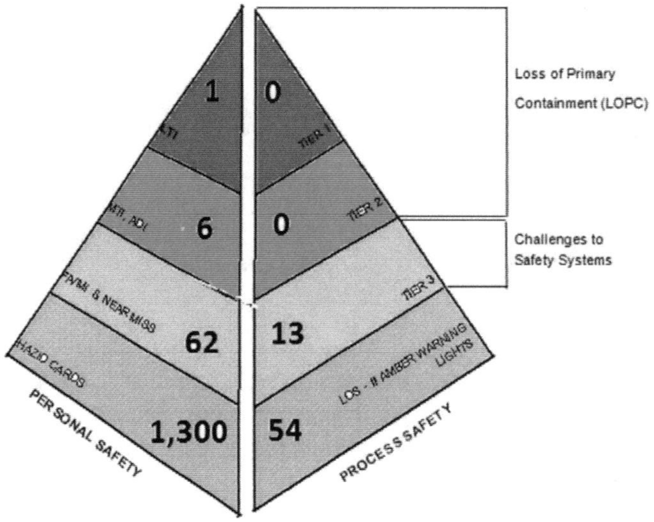

15

Mindfulness

- In a constant state of unease, concerned about complacency
- Sensitive to Operations, high situational awareness
- Resilient in event of setback
- Deferring to expertise
- Being alert to warning signs
  - Decisions on single point data
  - Abnormal process conditions
  - Intermittent changes
  - The feeling of operating blind and being out of control
- Proving it is safe, not unsafe

PTTEP

Summary

- PTTEP Australasia transformation since August 2009
  - management culture, operational capability, safety processes, environmental systems
- Emphasis is on both personal and process safety
- MAE prevention includes Line of Sight on critical controls
- Safety Critical Elements have associated Performance Standards
- Case to Operate required if controls or Performance Standards are compromised
- Performance metrics defined for personal and process safety
- Culture of "mindfulness" applicable to both personal and process safety

Process Safety and Safety activities in Mitsubishi Chemical Group

Masaki Nakagawa

Mitsubishi Chemical Corporation
Process Safety Technology Department
Environment, Safety and Quality Department
1-1, Marunouchi 1-chome, Chiyoda-ku
Tokyo 100-8251, JAPAN

MITSUBISHI CHEMICAL GROUP
SCIENCE AND TECHNOLOGY RESEARCH CENTER, INC.

Outline

1. Overview of Mitsubishi Chemical Holdings
2. Motivation
3. Safety Review (SR) activity
4. Risk evaluation method
5. Promoting discussion
6. Conclusion

THE KAITEKI COMPANY

Organization

20140512

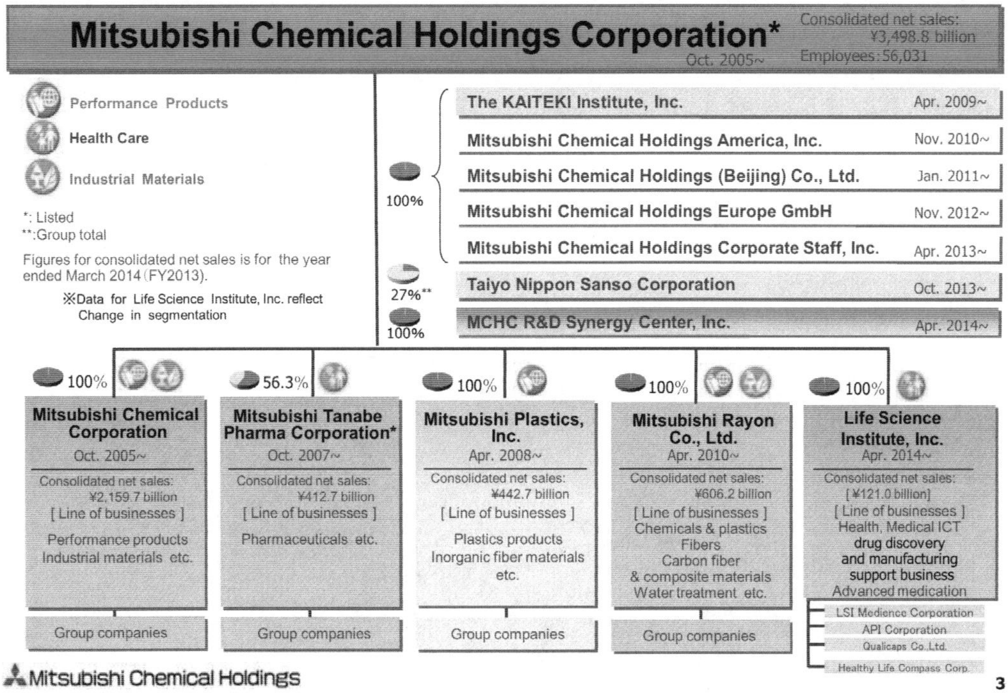

Mitsubishi Chemical Holdings

3

Outline

Business through Innovation

1. Overview of Mitsubishi Chemical Holdings
2. Motivation
3. Safety Review (SR) activity
4. Risk evaluation method
5. Promoting discussion
6. Conclusion

Motivation

Recently, the number of accidents which occur owing to the staff and operator's deficient knowledge or the shortage of experience is increasing because operating staff's opportunity to experience troubles is decreasing and veteran operator's retirement increases.
Therefore, making the mechanism of sharing the knowledge and the perspective from veteran staffs to young staffs is called for.
To share the knowledge, discussion is very effective.
This study is intended to improve knowledge transfer from veteran staffs to young staffs by promoting discussion.

Outline

1. Overview of Mitsubishi Chemical Holdings
2. Motivation
3. Safety Review (SR) activity
4. Risk evaluation method
5. Promoting discussion
6. Conclusion

Safety Review (SR) Activity

Business through Innovation

SR activities are joined not only by staffs who are familiar with routine operation, but also by staffs who belong to other departments.

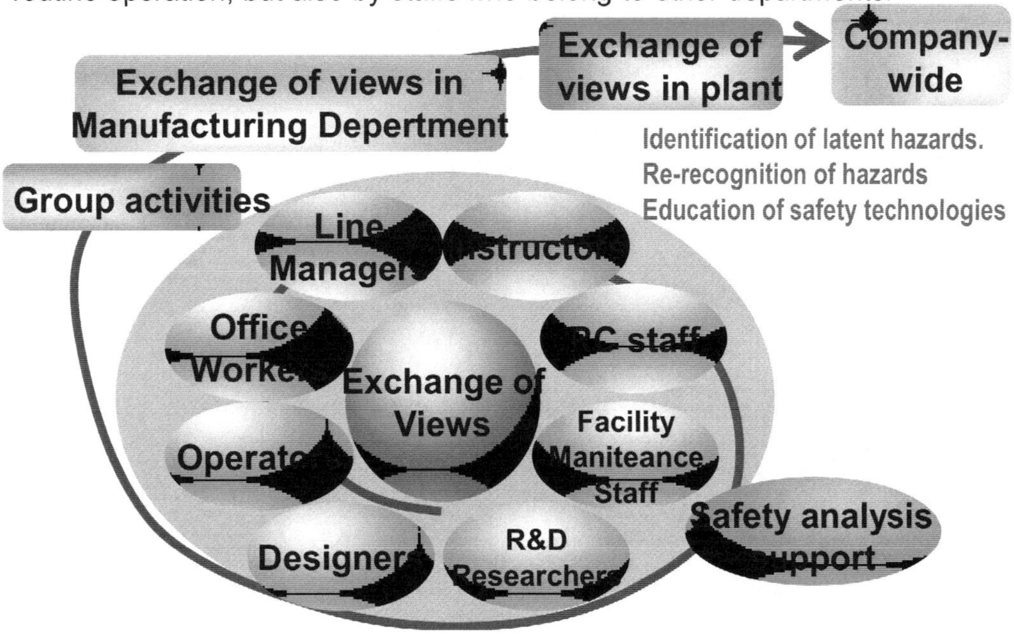

SR instructor

Business through Innovation

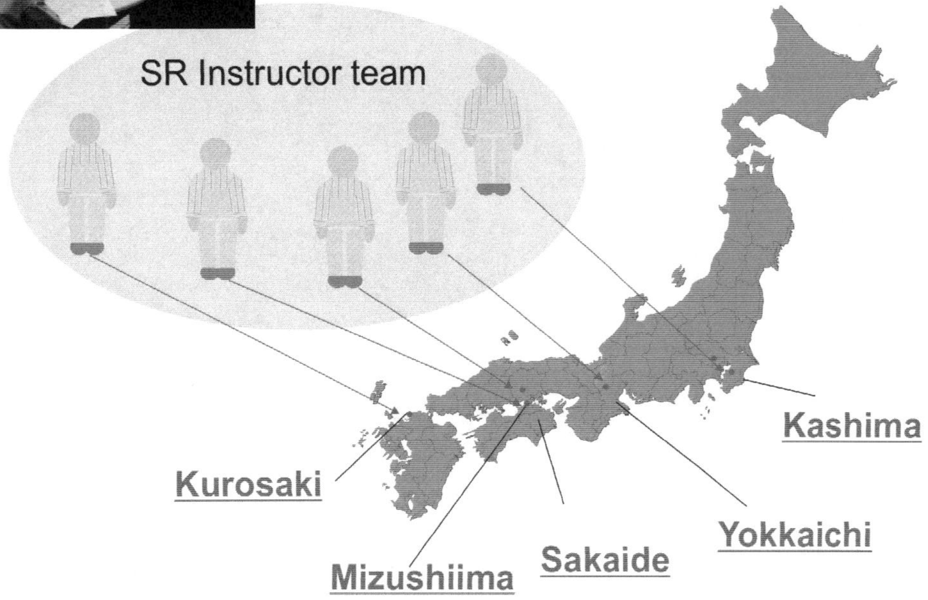

Education

Business through Innovation

<Process Safety Education program>

The safety education class in session

Basic course
Classroom lecture
20-30 attendee (per time, at a plant site)
Middle-tire engineers
12 hours

< Chemical Process Safety Engineer (CPSE) program >

Training for CPSE
conducted by visiting instructor

Advanced course
Small group seminars
5-6 attendees (per time, at a plant site)
Chosen engineers

Outline

Business through Innovation

1. Overview of Mitsubishi Chemical Holdings
2. Motivation
3. Safety Review (SR) activity
4. Risk evaluation method
   - ➡ HAZchart analysis
   - Case study
     - Original flow
     - Modified flow 1 (Minimal cut set)
     - Modified flow 2 (Common cause)
     - Modified flow 3 (Redundant system)
5. Promoting discussion
6. Conclusion

Risk evaluation method

HAZchart analysis is a tool for Identification of latent hazards.It standardizes procedure of accident scenario modeling. Accident scenario is modeled by querying three situations repeatedly. After that, a HAZchart is transferred to Fault Tree and the probability of accident can be calculated easily. HAZchart analysis enables us to consider the worst case scenario.

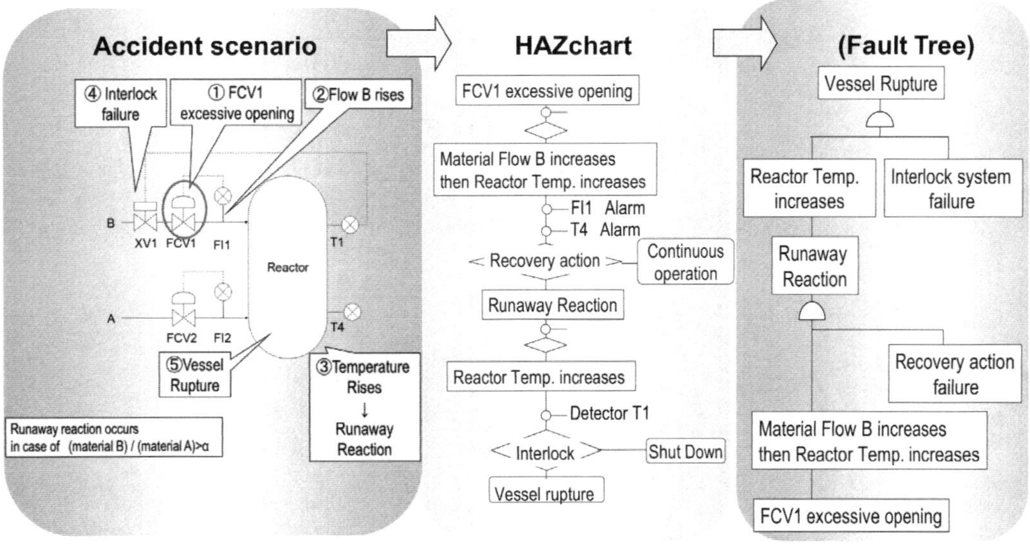

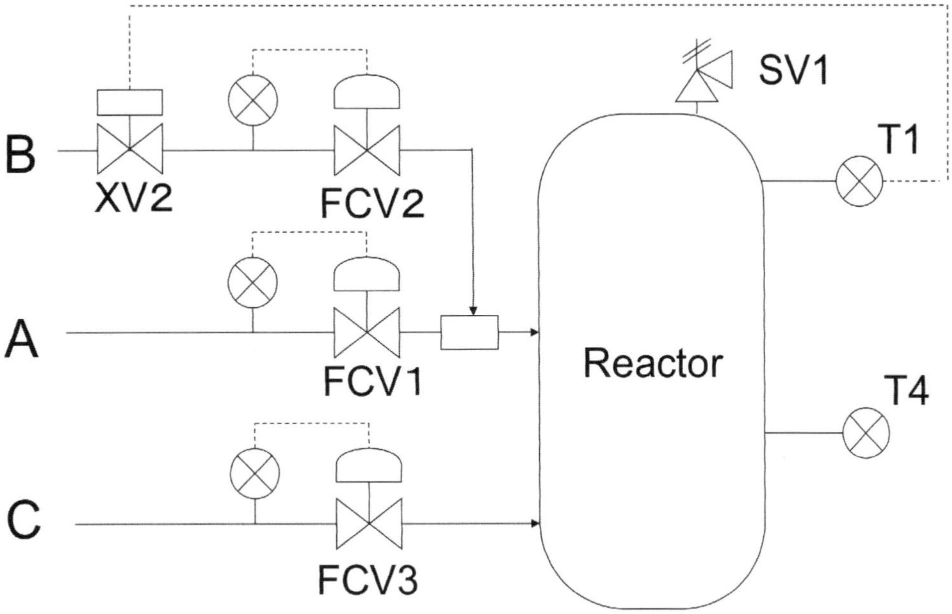

Runaway reaction occurs
in case of (material B) / (material A)> α

Procedure for HAZchart

-Selection of the fault mode of each equipment-

Selection of the each fault mode of the equipment which can potentially affect the state of the process fluid

Category of equipment	Mode of malfunction
Control valve	Excessive opening
	Too small opening
Shut off valve	Close even in normal condition
	open even in normal condition

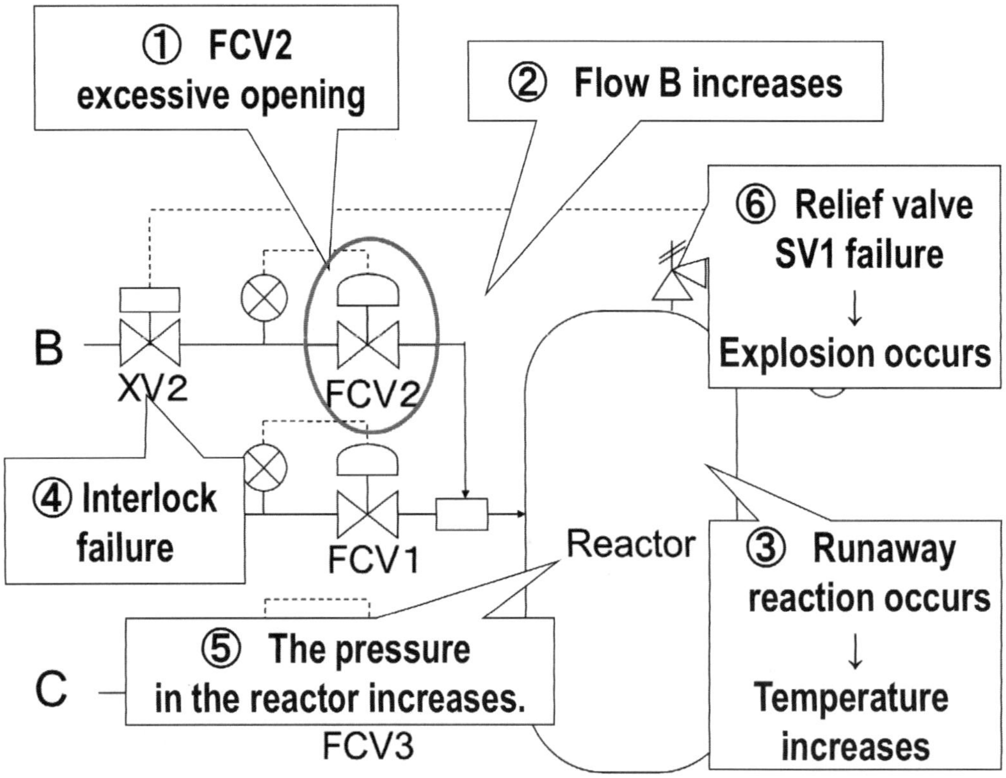

Modelling HAZchart

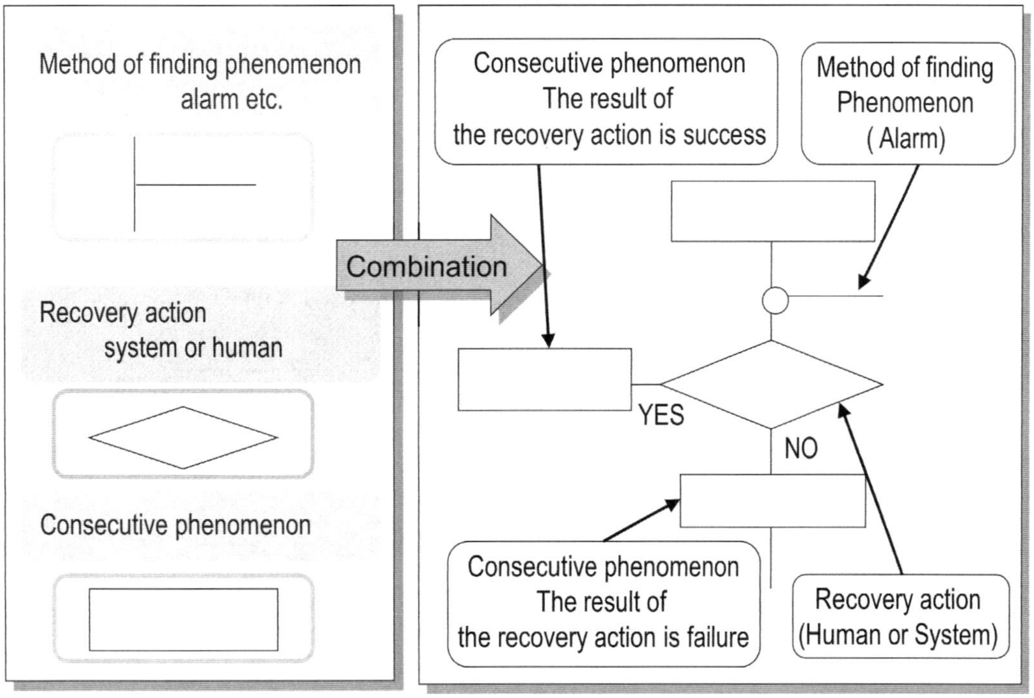

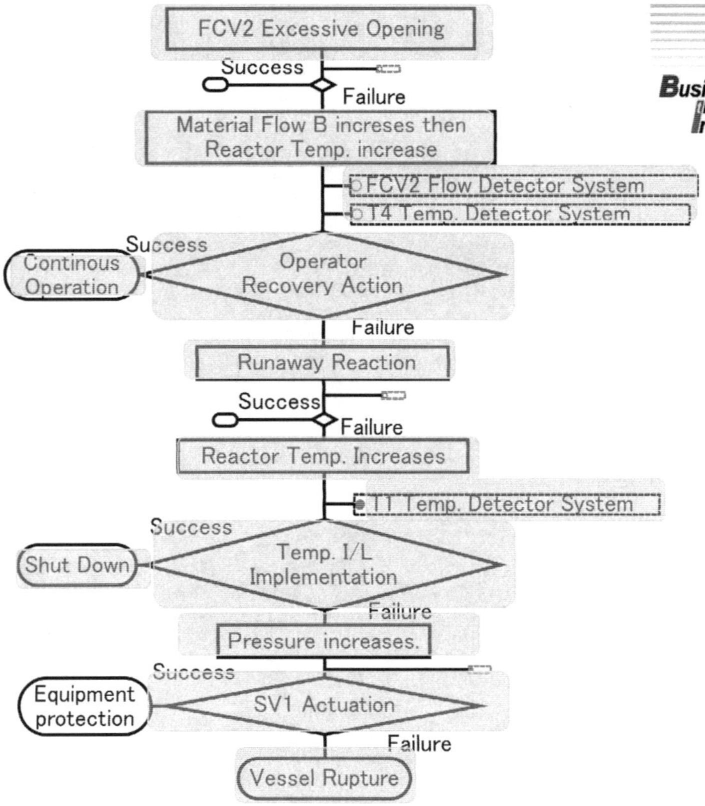

Correspondence of Protection Layer with HAZchart

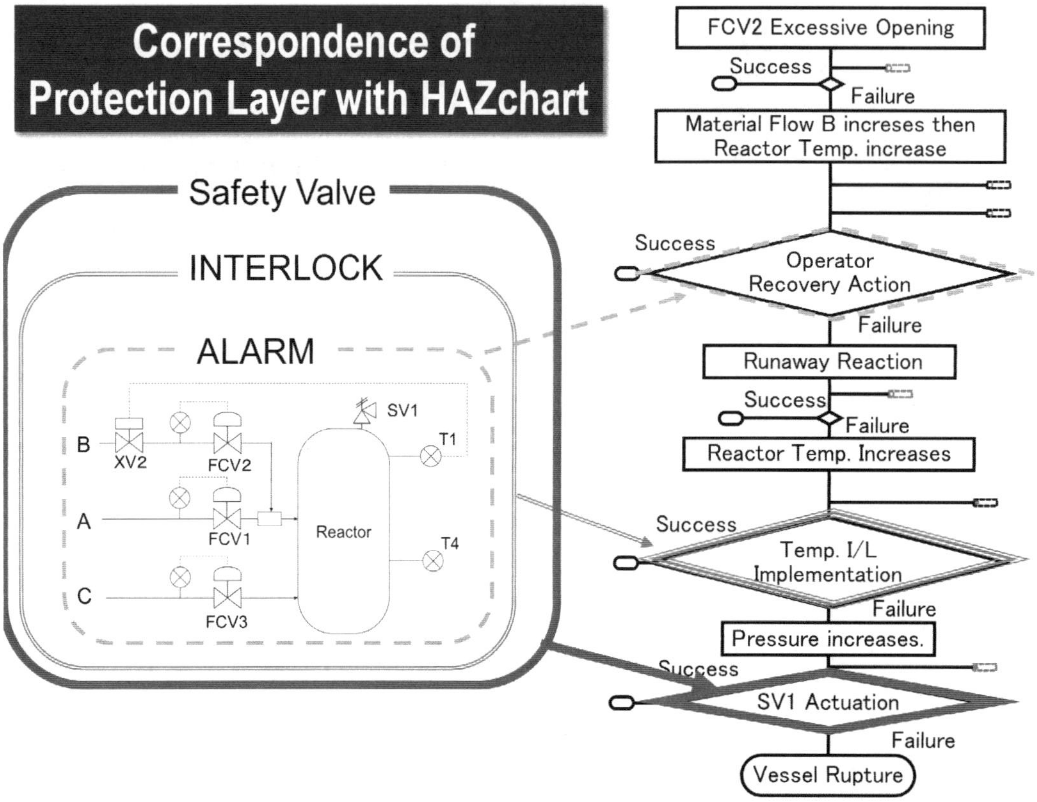

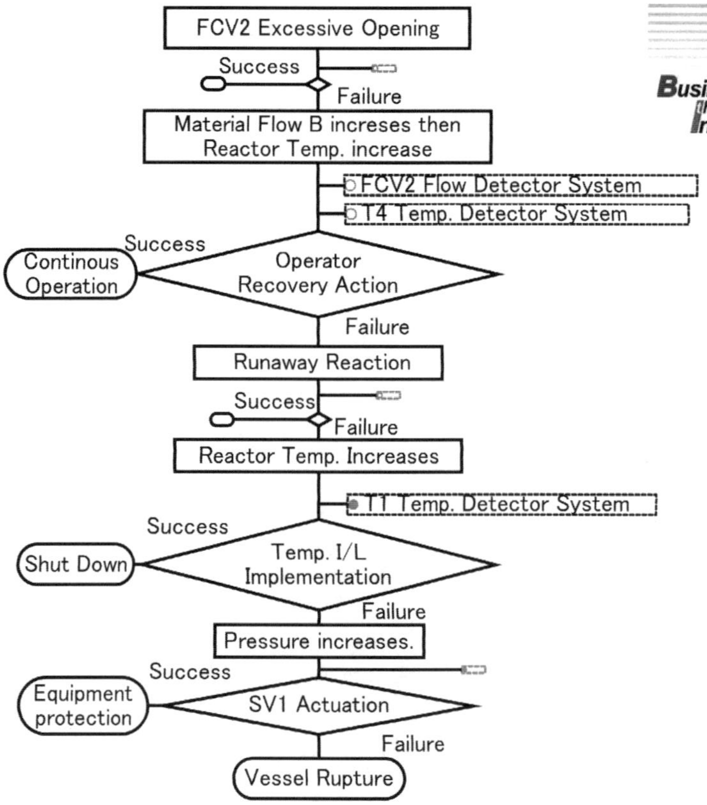

Business through Innovation

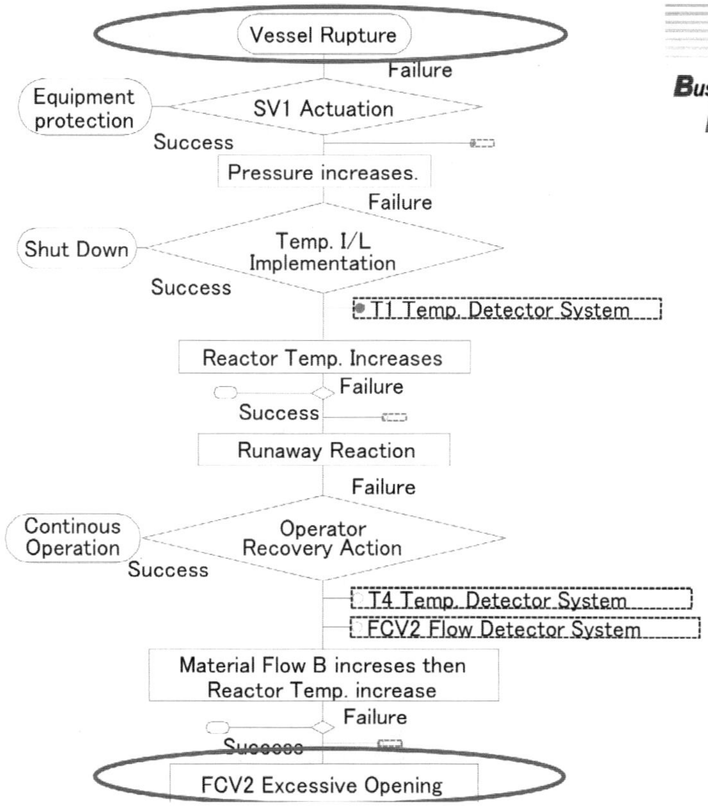
Vessel Rupture
Failure
Equipment protection
SV1 Actuation
Success
Pressure increases.
Failure
Shut Down
Temp. I/L Implementation
Success
T1 Temp. Detector System
Reactor Temp. Increases
Failure
Success
Runaway Reaction
Failure
Continous Operation
Operator Recovery Action
Success
T4 Temp. Detector System
FCV2 Flow Detector System
Material Flow B increses then Reactor Temp. increase
Failure
Success
FCV2 Excessive Opening

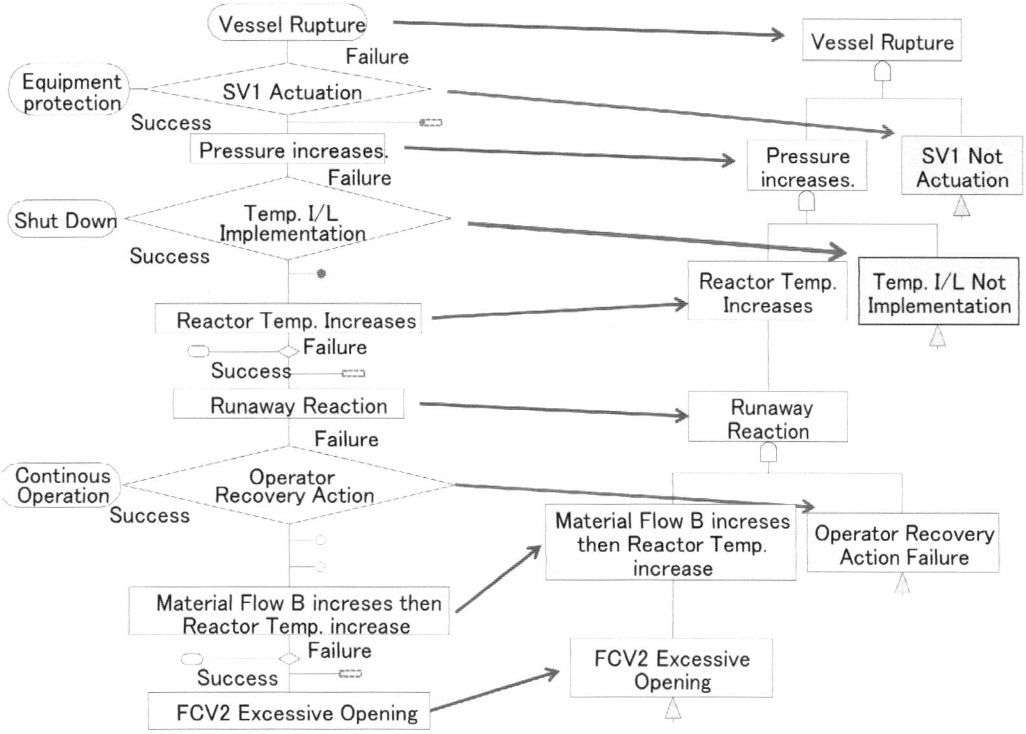
Vessel Rupture
Failure
Equipment protection
SV1 Actuation
Success
Pressure increases.
Failure
Shut Down
Temp. I/L Implementation
Success
Reactor Temp. Increases
Failure
Success
Runaway Reaction
Failure
Continous Operation
Operator Recovery Action
Success
Material Flow B increses then Reactor Temp. increase
Failure
Success
FCV2 Excessive Opening
Vessel Rupture
Pressure increases.
SV1 Not Actuation
Reactor Temp. Increases
Temp. I/L Not Implementation
Runaway Reaction
Material Flow B increses then Reactor Temp. increase
Operator Recovery Action Failure
FCV2 Excessive Opening

PHA_Organizer

HAZchart analysis method is packaged as a computer program that can reduce the time required for analysis.

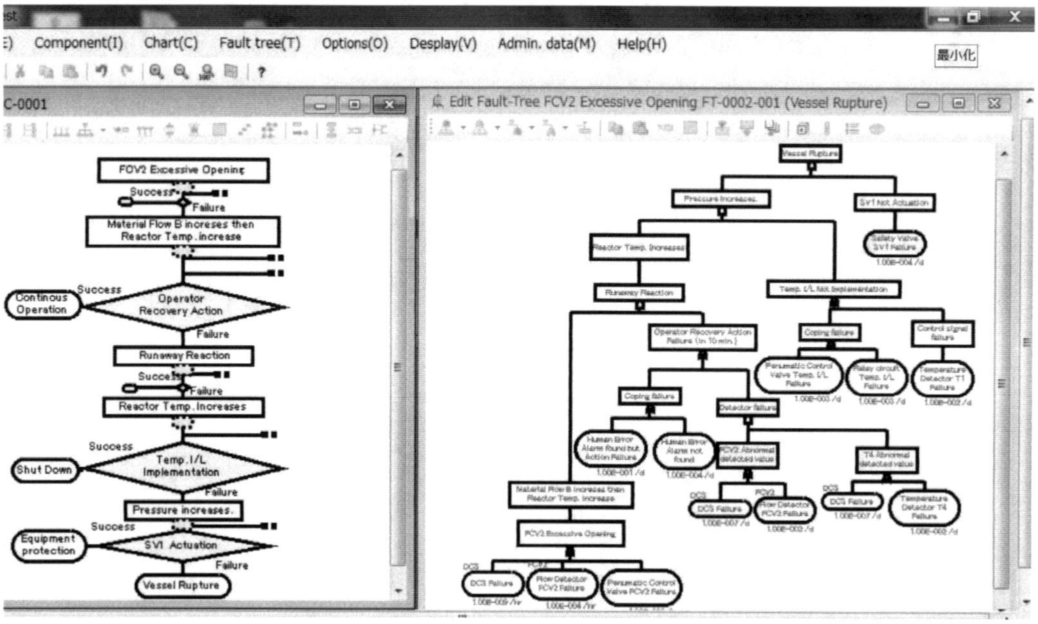

Fault Tree

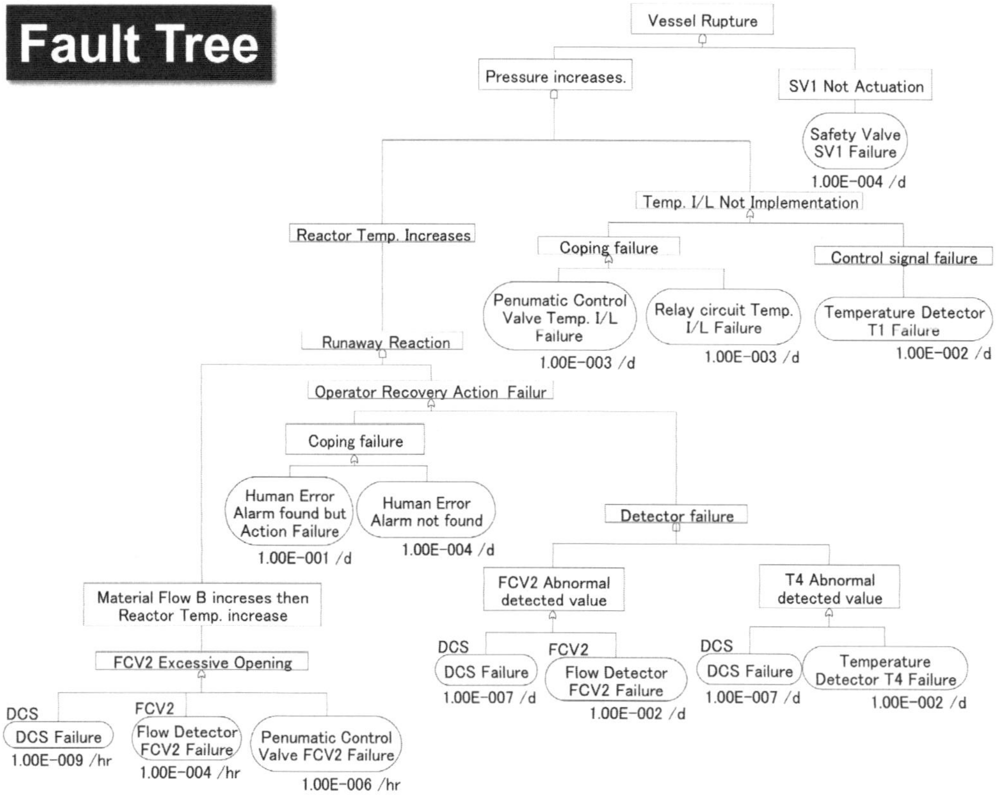

Risk Matrix

Business through Innovation

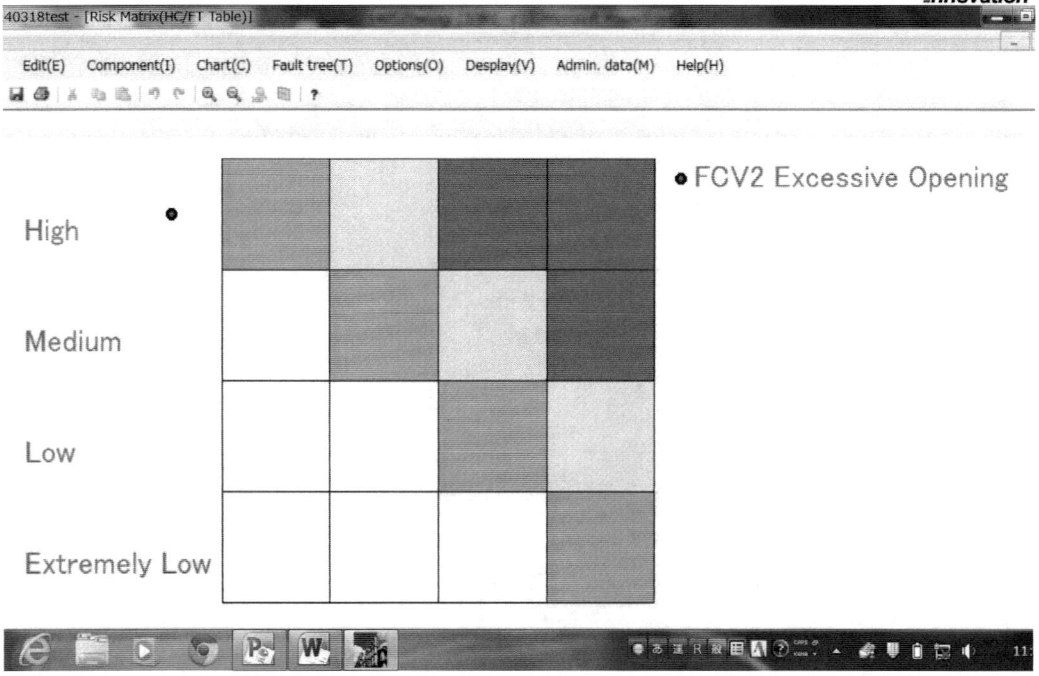

Outline

Business through Innovation

1. Overview of Mitsubishi Chemical Holdings
2. Motivation
3. Safety Review (SR) activity
4. Risk evaluation method
   - HAZchart analysis
   - ➡ Case study
     - Original flow
     - Modified flow 1 (Minimal cut set)
     - Modified flow 2 (Common cause)
     - Modified flow 3 (Redundant system)
5. Promoting discussion
6. Conclusion

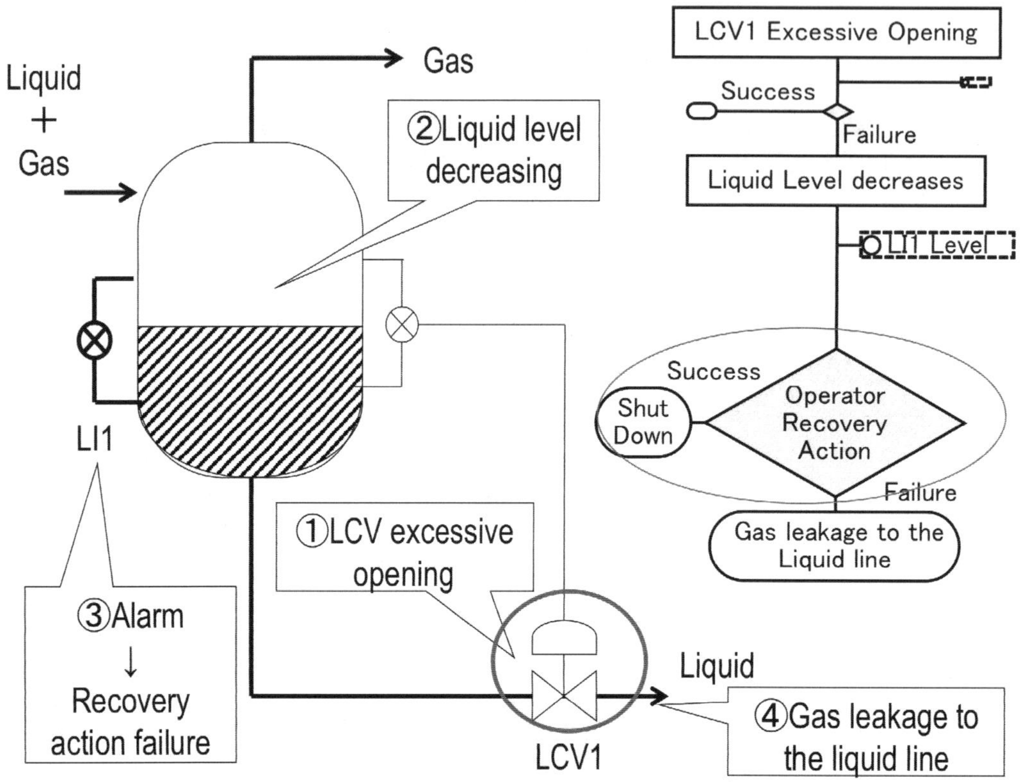
Liquid
+
Gas
Gas
②Liquid level decreasing
LI1
①LCV excessive opening
③Alarm
↓
Recovery action failure
LCV1
Liquid
④Gas leakage to the liquid line
LCV1 Excessive Opening
Success
Failure
Liquid Level decreases
LI1 Level
Success
Shut Down
Operator Recovery Action
Failure
Gas leakage to the Liquid line

Fault Tree

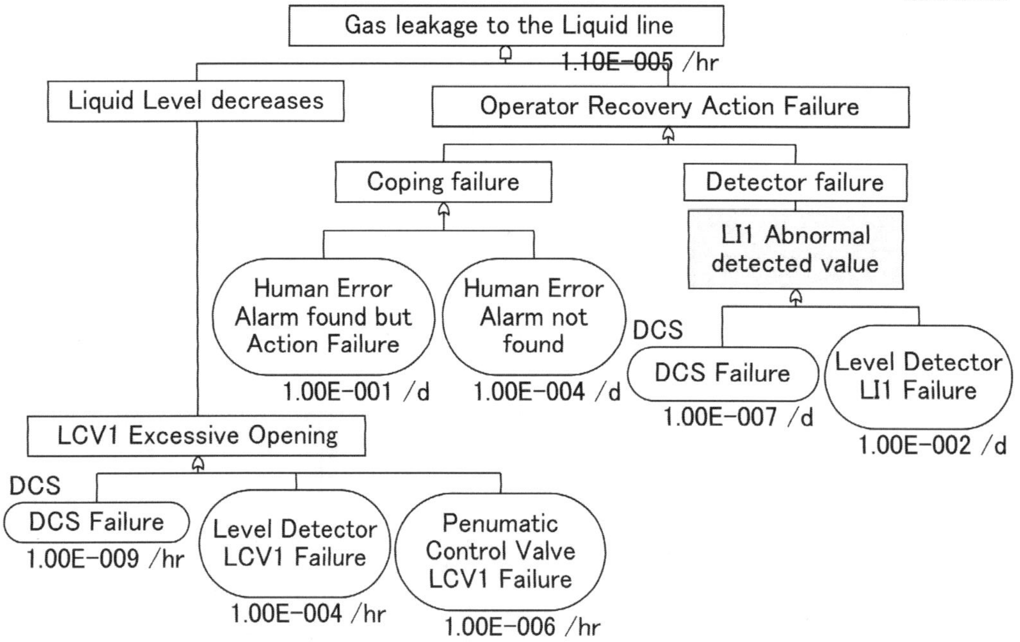
Gas leakage to the Liquid line
1.10E-005 /hr
Liquid Level decreases
Operator Recovery Action Failure
Coping failure
Detector failure
LI1 Abnormal detected value
Human Error Alarm found but Action Failure
1.00E-001 /d
Human Error Alarm not found
1.00E-004 /d
DCS
DCS Failure
1.00E-007 /d
Level Detector LI1 Failure
1.00E-002 /d
LCV1 Excessive Opening
DCS
DCS Failure
1.00E-009 /hr
Level Detector LCV1 Failure
1.00E-004 /hr
Penumatic Control Valve LCV1 Failure
1.00E-006 /hr

Risk Matrix

Business through Innovation

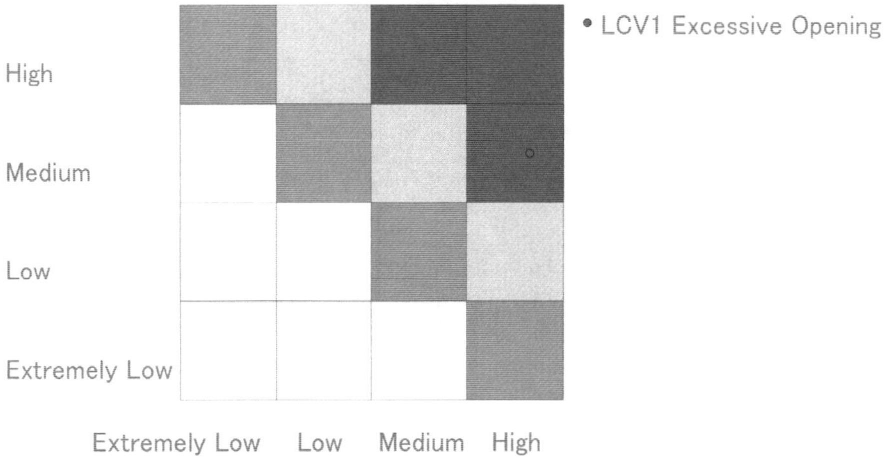

Outline

Business through Innovation

1. Overview of Mitsubishi Chemical Holdings
2. Motivation
3. Safety Review (SR) activity
4. Risk evaluation method
   - HAZchart analysis
   - Case study
     - Original flow
     - ➡ Modified flow 1 (Minimal cut set)
     - Modified flow 2 (Common cause)
     - Modified flow 3 (Redundant system)
5. Promoting discussion
6. Conclusion

Minimal Cut Set

A cut set is a combination of basic phenomena which generates the top phenomenon of FT. A minimal cut set is made required to generate a top phenomenon of a system, and is sufficient cut set.

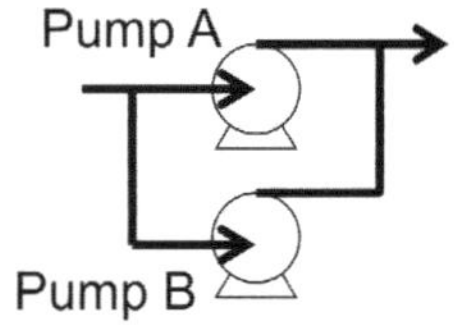

The power supply of the pumps is taken from the same power source.

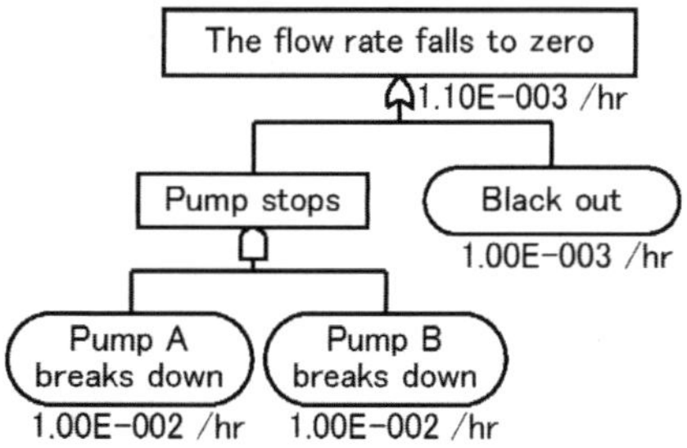

The minimal cut set was often expressed as a set.

(Black out)
(Pump A break down, Pump B break down)

Minimal Cut Set

A cut set is a combination of basic phenomena which generates the top phenomenon of FT. A minimal cut set is made required to generate a top phenomenon of a system, and is sufficient cut set.

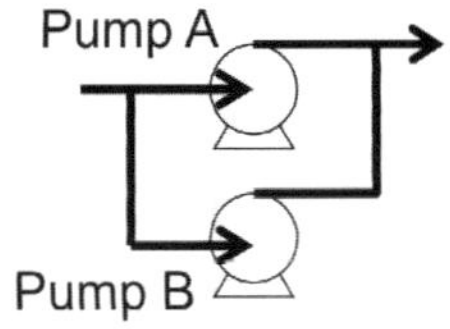

The power supply of the pumps is taken from the same power source.

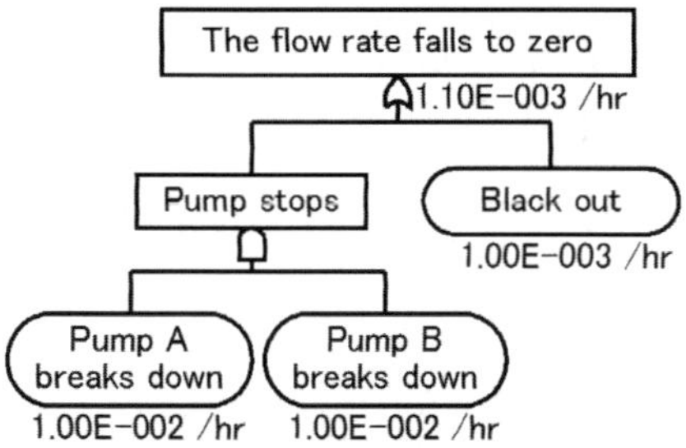

The function of displaying minimal cut set relationship on the FT

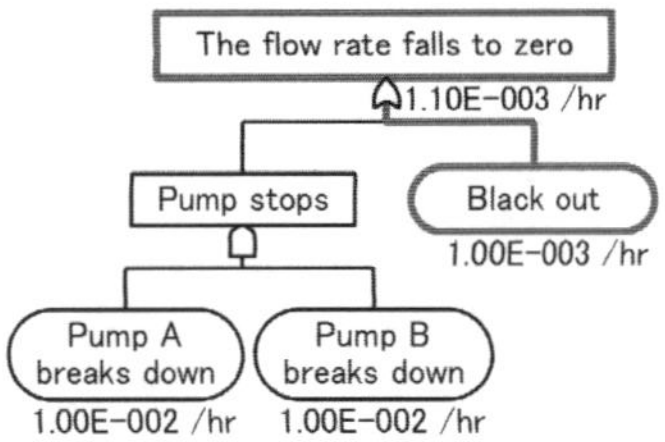

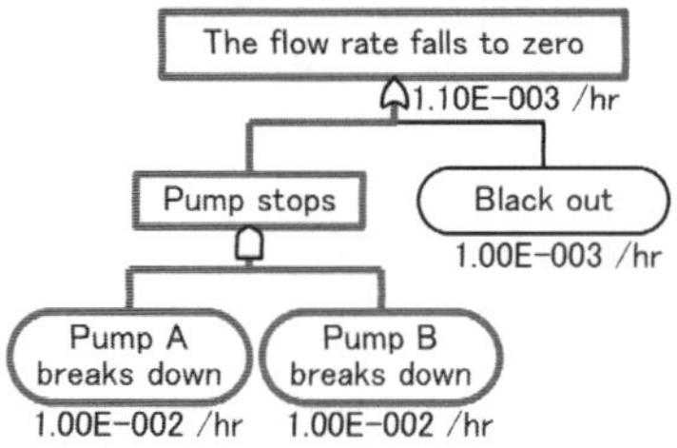

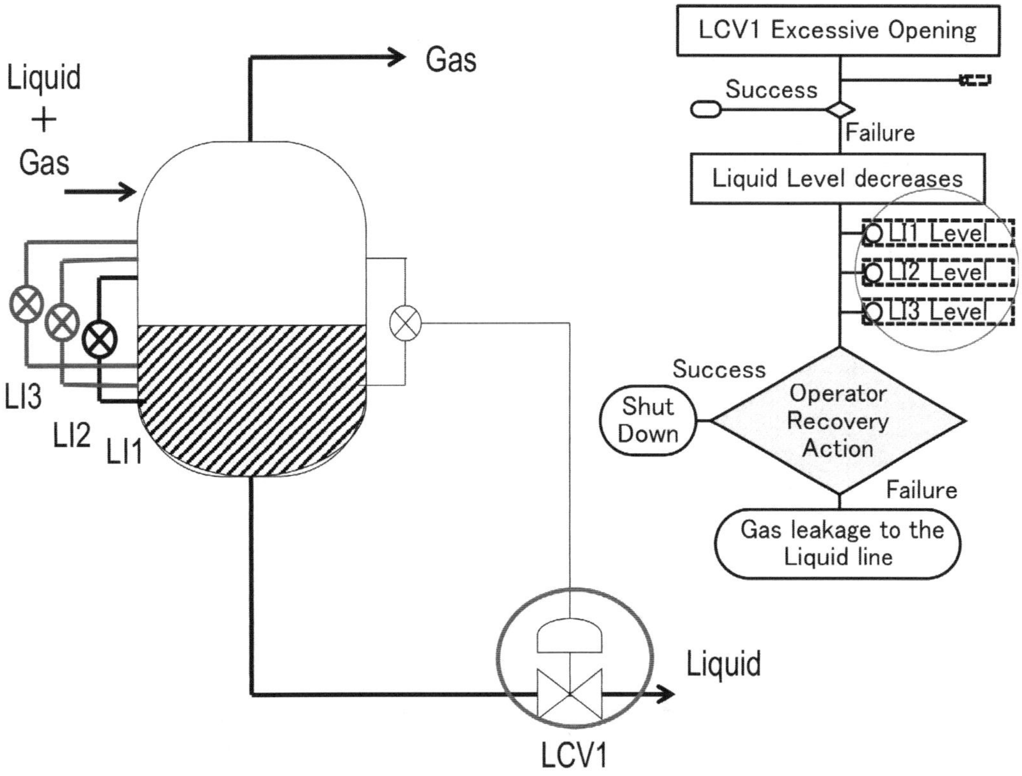

Risk Matrix

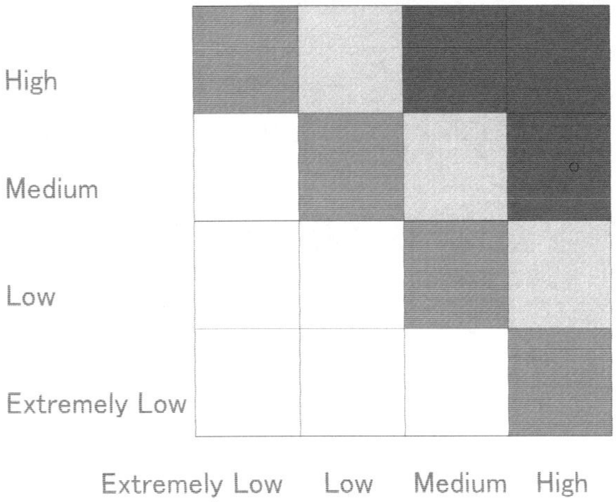

- LCV1 Excessive Opening

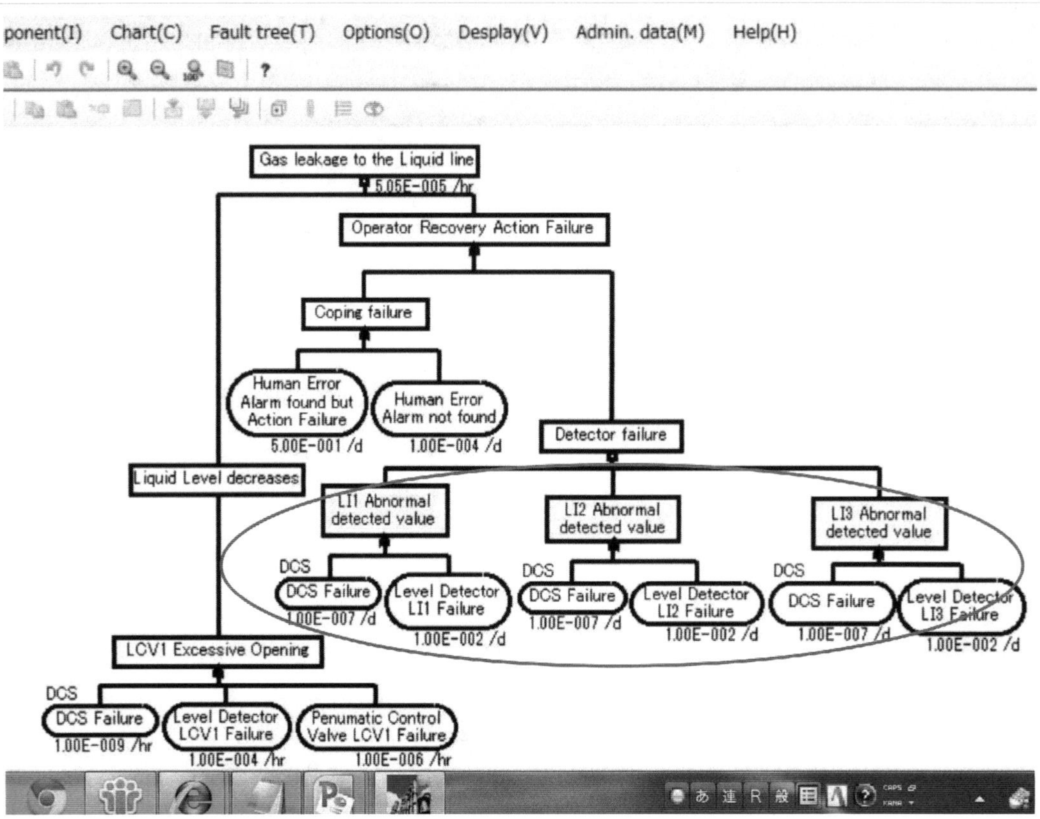
ponent(I) Chart(C) Fault tree(T) Options(O) Desplay(V) Admin. data(M) Help(H)
Gas leakage to the Liquid line
5.05E-005 /hr
Operator Recovery Action Failure
Coping failure
Human Error Alarm found but Action Failure
5.00E-001 /d
Human Error Alarm not found
1.00E-004 /d
Detector failure
Liquid Level decreases
LI1 Abnormal detected value
LI2 Abnormal detected value
LI3 Abnormal detected value
DCS
DCS Failure
1.00E-007 /d
Level Detector LI1 Failure
1.00E-002 /d
Level Detector LI2 Failure
Level Detector LI3 Failure
LCV1 Excessive Opening
1.00E-009 /hr
Level Detector LCV1 Failure
1.00E-004 /hr
Penumatic Control Valve LCV1 Failure
1.00E-006 /hr

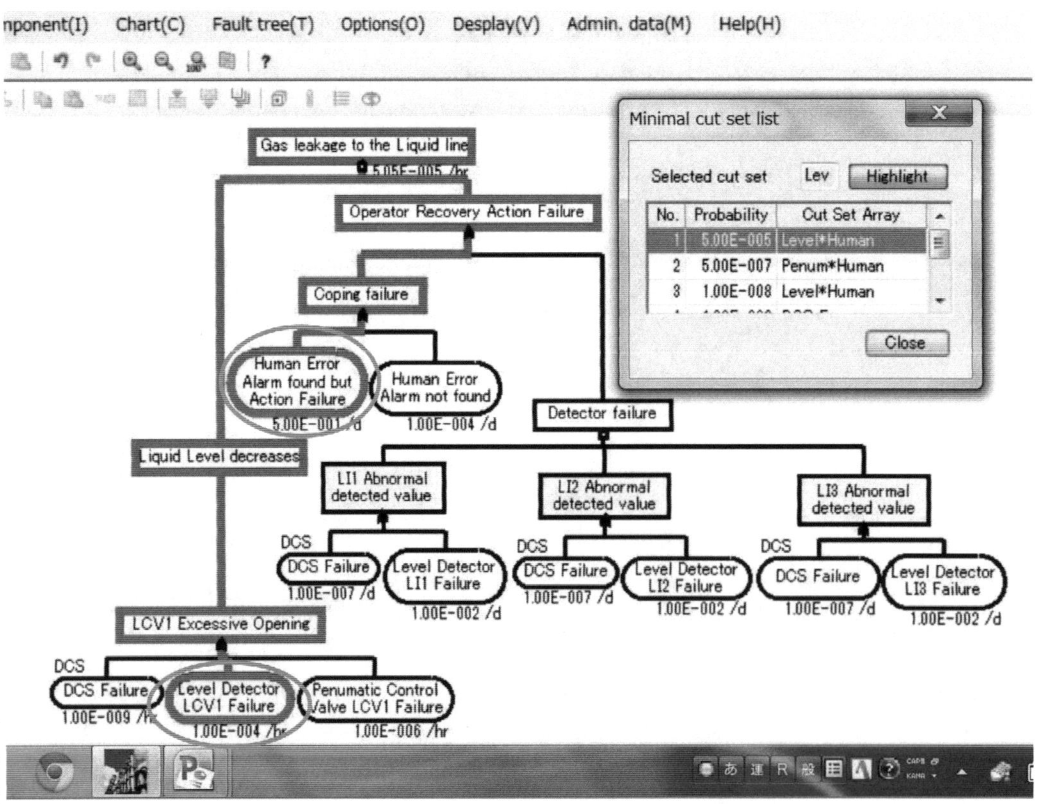
nponent(I) Chart(C) Fault tree(T) Options(O) Desplay(V) Admin. data(M) Help(H)
Minimal cut set list
Selected cut set
Lev
Highlight
No.
Probability
Cut Set Array
1 5.00E-005 Level*Human
2 5.00E-007 Penum*Human
3 1.00E-008 Level*Human
Close
Gas leakage to the Liquid line
Operator Recovery Action Failure
Coping failure
Human Error Alarm found but Action Failure
5.00E-001 /d
Human Error Alarm not found
1.00E-004 /d
Detector failure
Liquid Level decreases
LI1 Abnormal detected value
LI2 Abnormal detected value
LI3 Abnormal detected value
DCS
DCS Failure
1.00E-007 /d
Level Detector LI1 Failure
1.00E-002 /d
Level Detector LI2 Failure
Level Detector LI3 Failure
LCV1 Excessive Opening
1.00E-009 /hr
Level Detector LCV1 Failure
1.00E-004 /hr
Penumatic Control Valve LCV1 Failure
1.00E-006 /hr

Outline

Business through Innovation

1. Overview of Mitsubishi Chemical Holdings
2. Motivation
3. Safety Review (SR) activity
4. Risk evaluation method
   - HAZchart analysis
   - Case study
     - Original flow
     - Modified flow 1 (Minimal cut set)
     - ➡ Modified flow 2 (Common cause)
     - Modified flow 3 (Redundant system)
5. Promoting discussion
6. Conclusion

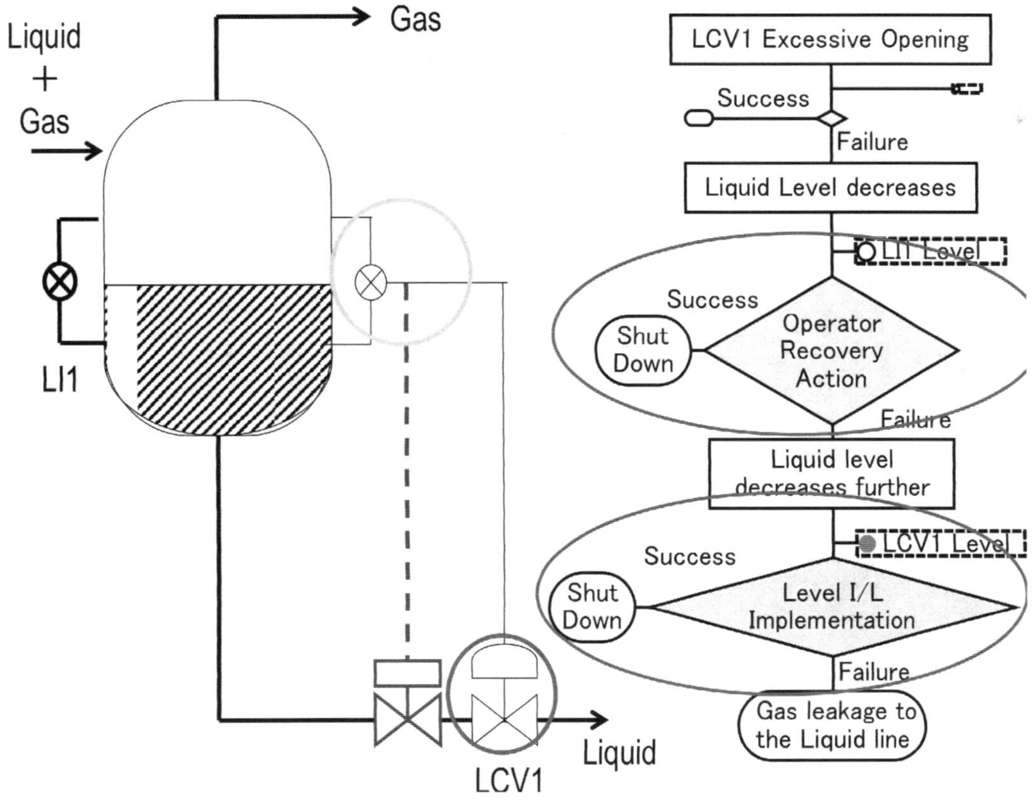

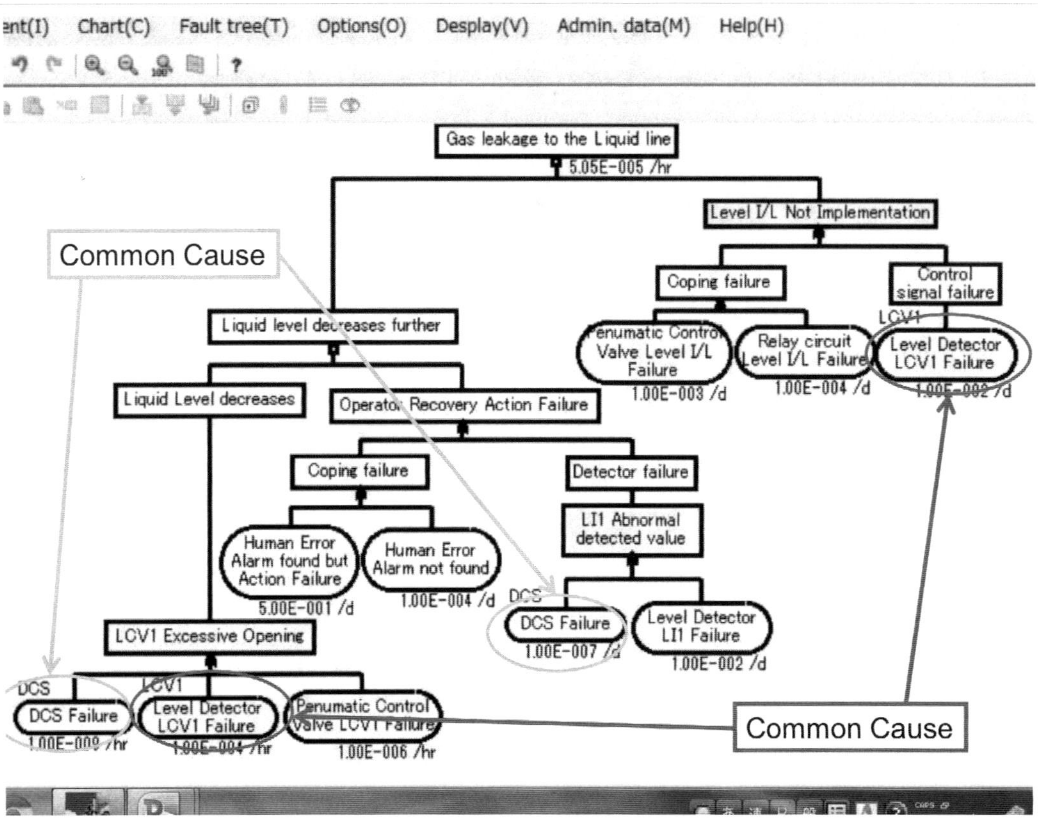
ent(I) Chart(C) Fault tree(T) Options(O) Desplay(V) Admin. data(M) Help(H)
Gas leakage to the Liquid line
5.05E-005 /hr
Level I/L Not Implementation
Common Cause
Coping failure
Control signal failure
Liquid level decreases further
LCV1
Penumatic Control Valve Level I/L Failure
Relay circuit Level I/L Failure
Level Detector LCV1 Failure
1.00E-003 /d
1.00E-004 /d
Liquid Level decreases
Operator Recovery Action Failure
Coping failure
Detector failure
LI1 Abnormal detected value
Human Error Alarm found but Action Failure
Human Error Alarm not found
5.00E-001 /d
1.00E-004 /d
DCS
DCS Failure
Level Detector LI1 Failure
1.00E-007 /d
1.00E-002 /d
LCV1 Excessive Opening
DCS
LCV1
DCS Failure
Level Detector LCV1 Failure
Penumatic Control Valve LCV1 Failure
1.00E-006 /hr
Common Cause

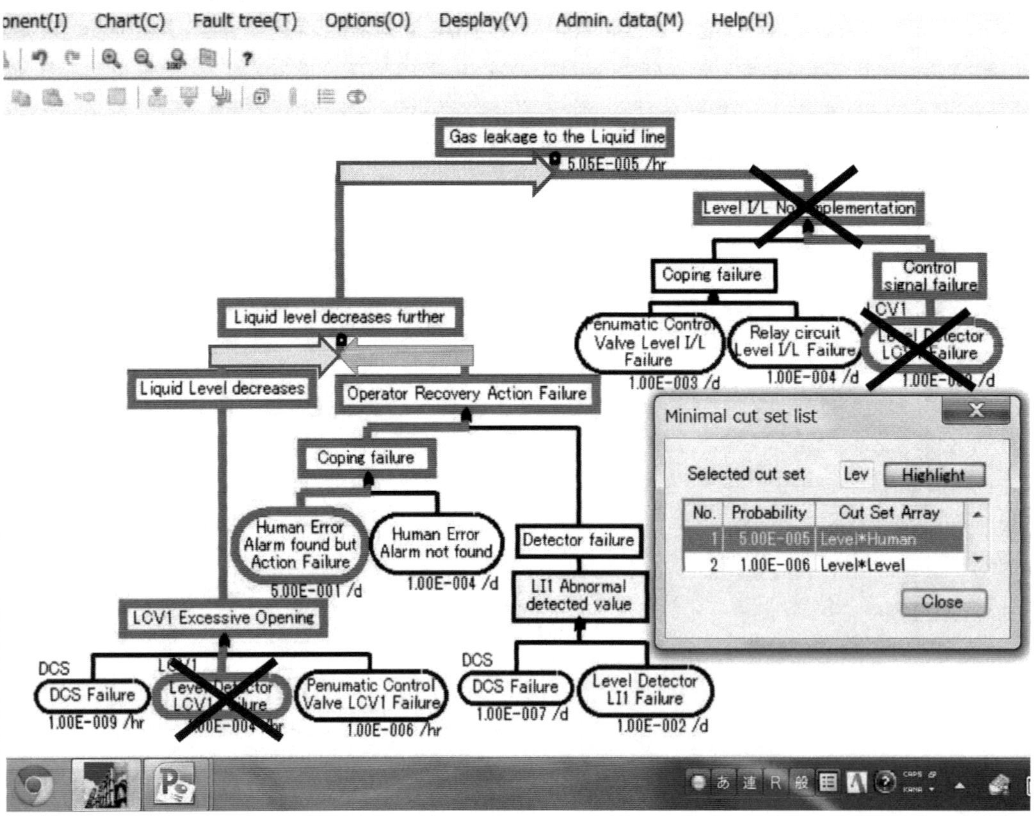
onent(I) Chart(C) Fault tree(T) Options(O) Desplay(V) Admin. data(M) Help(H)
Gas leakage to the Liquid line
5.05E-005 /hr
Coping failure
Control signal failure
LCV1
Liquid level decreases further
Penumatic Control Valve Level I/L Failure
Relay circuit Level I/L Failure
1.00E-003 /d
1.00E-004 /d
Liquid Level decreases
Operator Recovery Action Failure
Minimal cut set list
Selected cut set
Lev
Highlight
No.
Probability
Cut Set Array
1
5.00E-005
Level*Human
2
1.00E-006
Level*Level
Close
Coping failure
Human Error Alarm found but Action Failure
Human Error Alarm not found
Detector failure
5.00E-001 /d
1.00E-004 /d
LI1 Abnormal detected value
LCV1 Excessive Opening
DCS
DCS
DCS Failure
Penumatic Control Valve LCV1 Failure
DCS Failure
Level Detector LI1 Failure
1.00E-009 /hr
1.00E-006 /hr
1.00E-007 /d
1.00E-002 /d

Risk Matrix

Business through Innovation

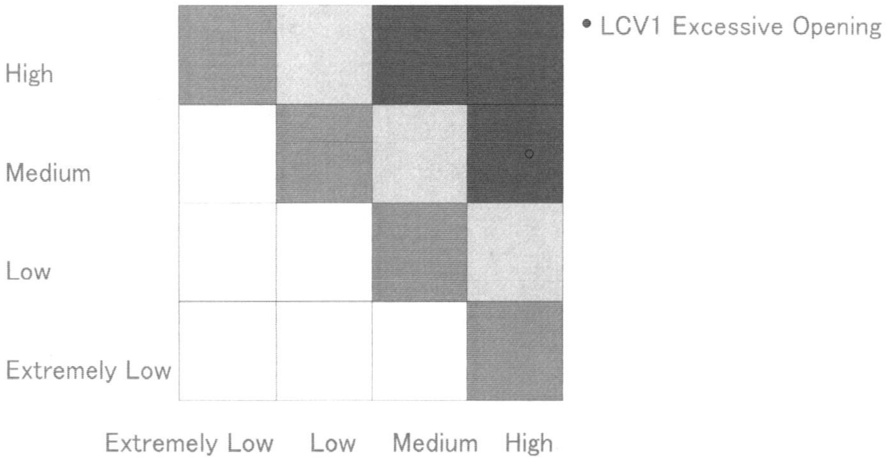

Outline

Business through Innovation

1. Overview of Mitsubishi Chemical Holdings
2. Motivation
3. Safety Review (SR) activity
4. Risk evaluation method
   - HAZchart analysis
   - Case study
     - Original flow
     - Modified flow 1 (Minimal cut set)
     - Modified flow 2 (Common cause)
     - Modified flow 3 (Redundant system)
5. Promoting discussion
6. Conclusion

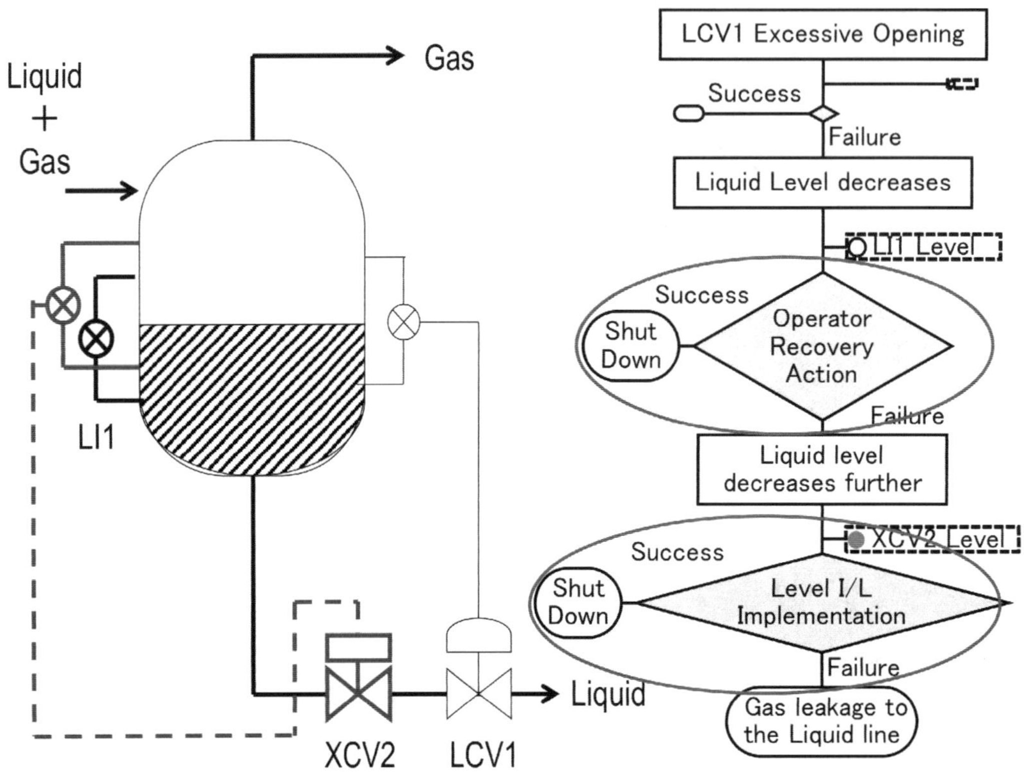
Liquid
+
Gas
Gas
LI1
XCV2
LCV1
Liquid
LCV1 Excessive Opening
Success
Failure
Liquid Level decreases
LI1 Level
Success
Shut Down
Operator Recovery Action
Failure
Liquid level decreases further
XCV2 Level
Success
Shut Down
Level I/L Implementation
Failure
Gas leakage to the Liquid line

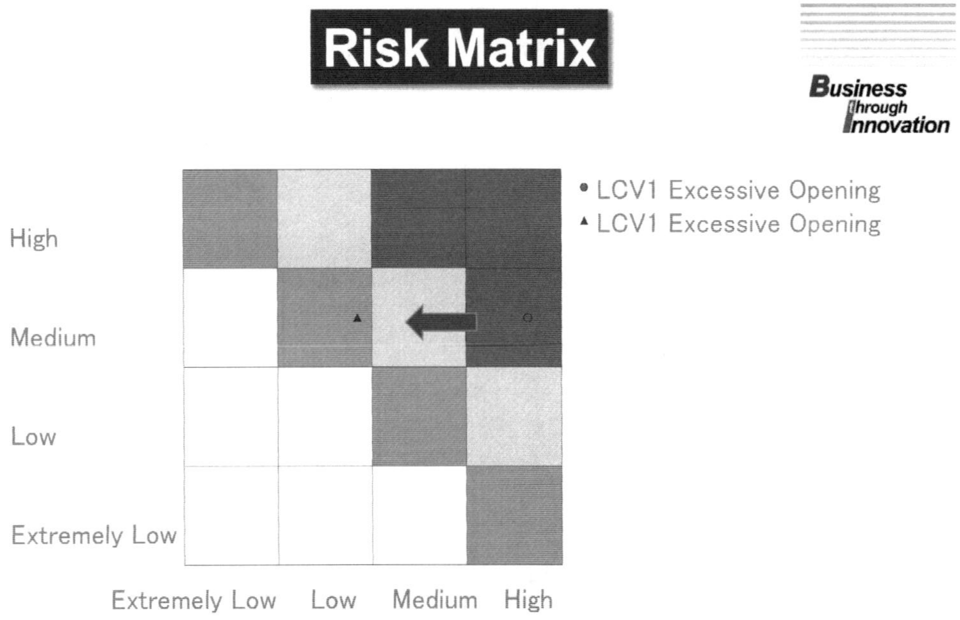
Risk Matrix
Business through Innovation
LCV1 Excessive Opening
LCV1 Excessive Opening
High
Medium
Low
Extremely Low
Extremely Low
Low
Medium
High

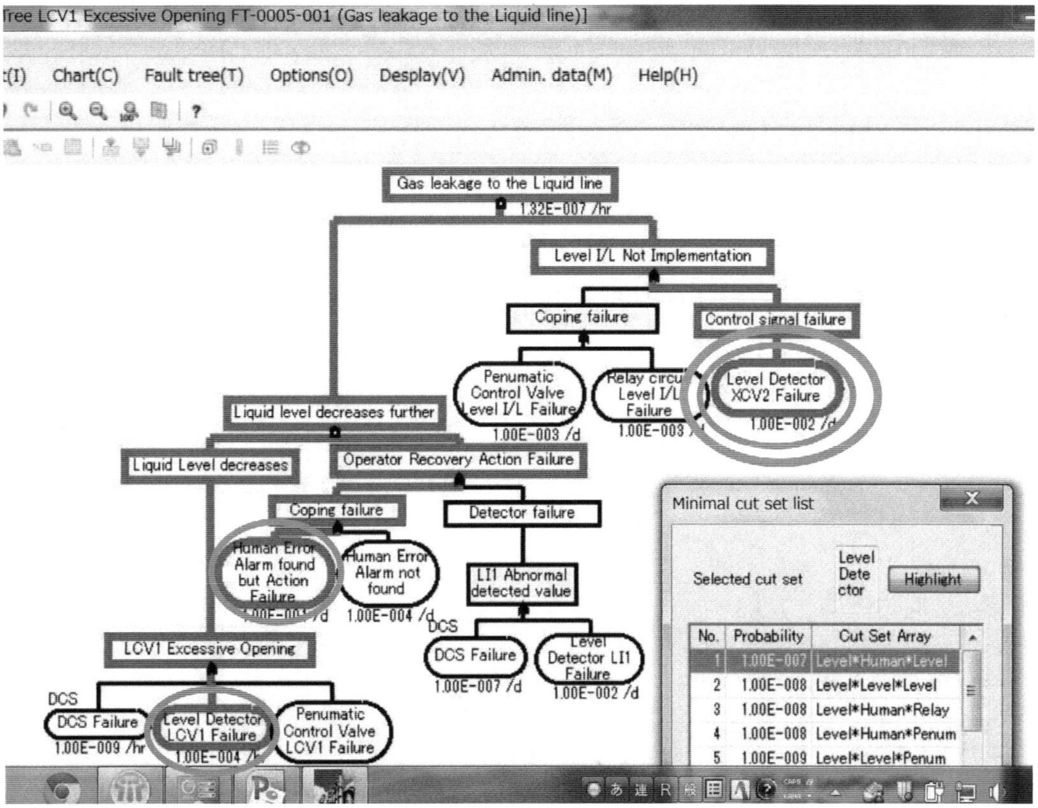
Tree LCV1 Excessive Opening FT-0005-001 (Gas leakage to the Liquid line)]
Chart(C) Fault tree(T) Options(O) Desplay(V) Admin. data(M) Help(H)
Gas leakage to the Liquid line
1.32E-007 /hr
Level I/L Not Implementation
Coping failure
Control signal failure
Penumatic Control Valve Level I/L Failure
1.00E-003 /d
Relay circuit Level I/L Failure
Level Detector XCV2 Failure
1.00E-002 /d
Liquid level decreases further
Liquid Level decreases
Operator Recovery Action Failure
Coping failure
Detector failure
Human Error Alarm found but Action Failure
Human Error Alarm not found
1.00E-004 /d
LI1 Abnormal detected value
DCS
DCS Failure
1.00E-007 /d
Level Detector LI1 Failure
1.00E-002 /d
LCV1 Excessive Opening
DCS
DCS Failure
1.00E-009 /hr
Level Detector LCV1 Failure
1.00E-004
Penumatic Control Valve LCV1 Failure
Minimal cut set list
Selected cut set
Level Dete ctor
Highlight
No. Probability Cut Set Array
1 1.00E-007 Level*Human*Level
2 1.00E-008 Level*Level*Level
3 1.00E-008 Level*Human*Relay
4 1.00E-008 Level*Human*Penum
5 1.00E-009 Level*Level*Penum

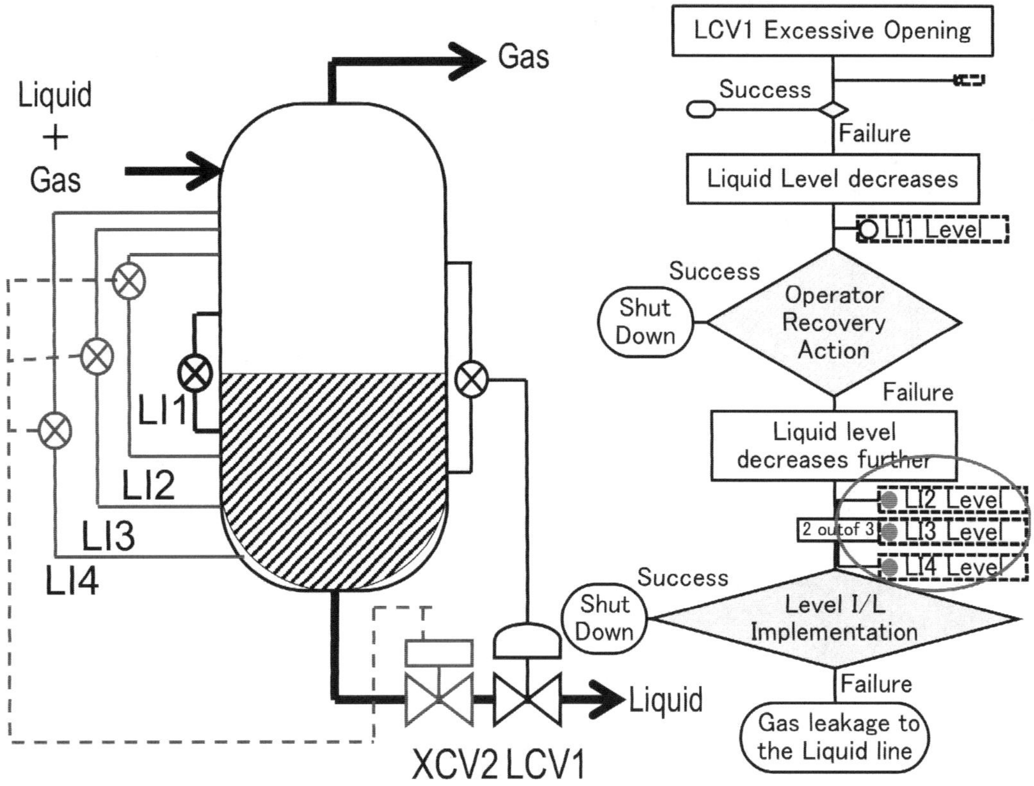
Gas
Liquid
+
Gas
LI1
LI2
LI3
LI4
Liquid
XCV2 LCV1
LCV1 Excessive Opening
Success
Failure
Liquid Level decreases
LI1 Level
Success
Shut Down
Operator Recovery Action
Failure
Liquid level decreases further
LI2 Level
2 out of 3
LI3 Level
LI4 Level
Success
Shut Down
Level I/L Implementation
Failure
Gas leakage to the Liquid line

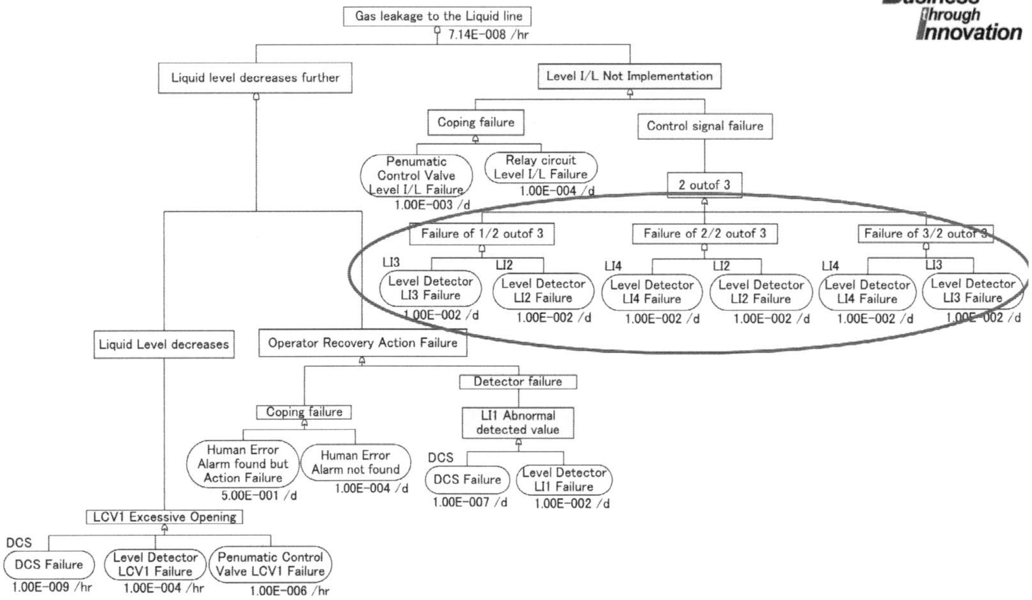

Gas leakage to the Liquid line
7.14E-008 /hr
Liquid level decreases further
Level I/L Not Implementation
Coping failure
Control signal failure
Penumatic Control Valve Level I/L Failure
1.00E-003 /d
Relay circuit Level I/L Failure
1.00E-004 /d
2 outof 3
Failure of 1/2 outof 3
Failure of 2/2 outof 3
Failure of 3/2 outof 3
LI3
LI2
LI4
LI2
LI4
LI3
Level Detector LI3 Failure
1.00E-002 /d
Level Detector LI2 Failure
1.00E-002 /d
Level Detector LI4 Failure
1.00E-002 /d
Level Detector LI2 Failure
1.00E-002 /d
Level Detector LI4 Failure
1.00E-002 /d
Level Detector LI3 Failure
1.00E-002 /d
Liquid Level decreases
Operator Recovery Action Failure
Detector failure
Coping failure
LI1 Abnormal detected value
Human Error Alarm found but Action Failure
5.00E-001 /d
Human Error Alarm not found
1.00E-004 /d
DCS
DCS Failure
1.00E-007 /d
Level Detector LI1 Failure
1.00E-002 /d
LCV1 Excessive Opening
DCS
DCS Failure
1.00E-009 /hr
Level Detector LCV1 Failure
1.00E-004 /hr
Penumatic Control Valve LCV1 Failure
1.00E-006 /hr

Risk Matrix

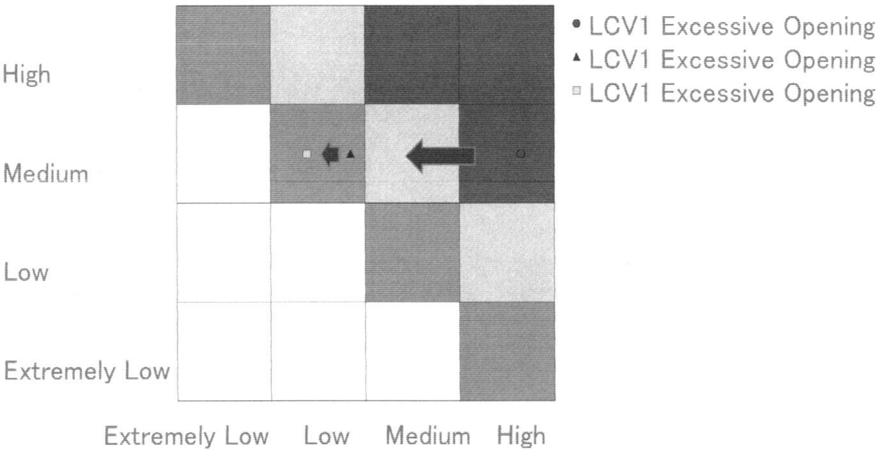

High
Medium
Low
Extremely Low
Extremely Low
Low
Medium
High
LCV1 Excessive Opening
LCV1 Excessive Opening
LCV1 Excessive Opening

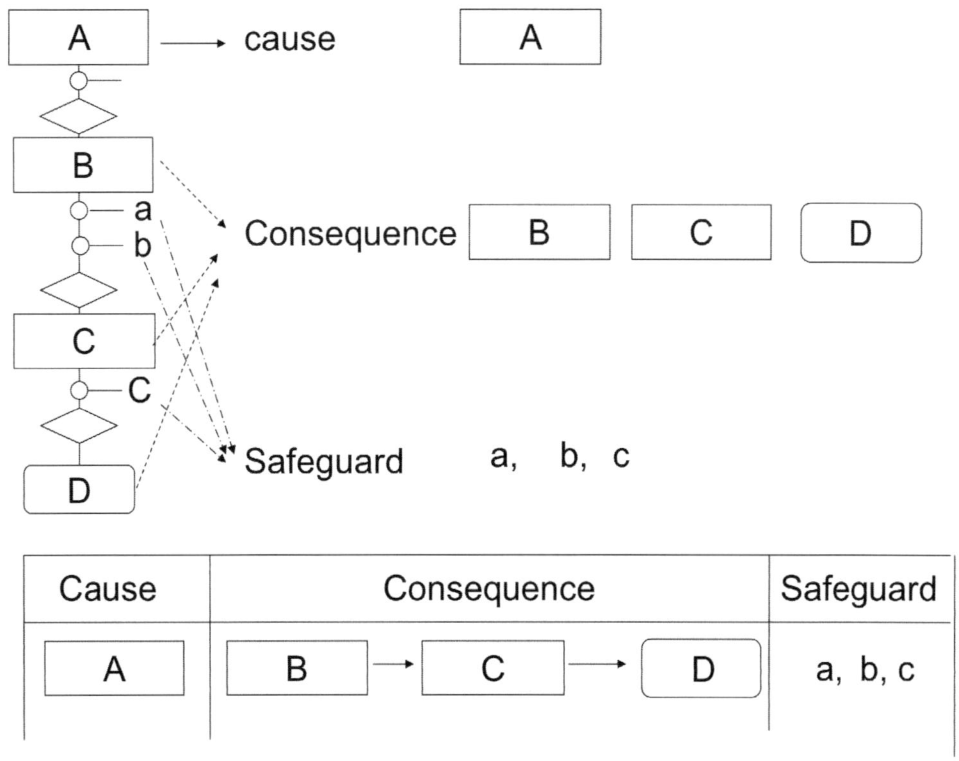
A
cause
A
B
a
b
Consequence
B
C
D
C
c
D
Safeguard
a, b, c
Cause
Consequence
Safeguard
A
B
C
D
a, b, c

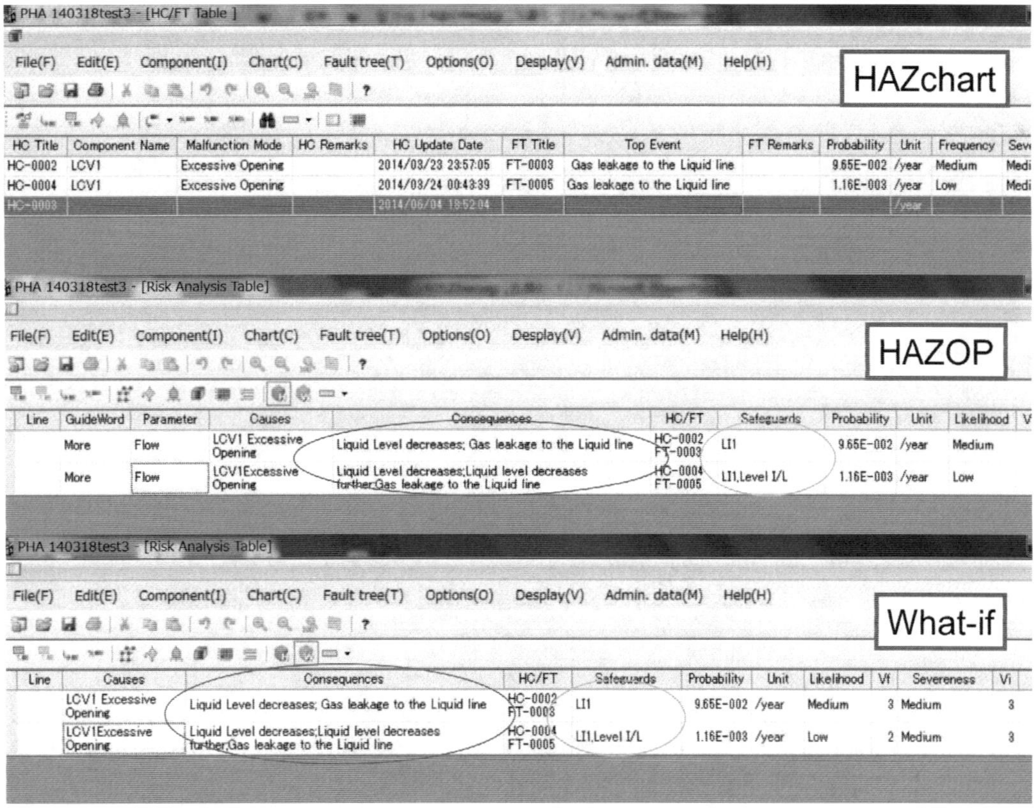
PHA 140318test3 - [HC/FT Table]
File(F) Edit(E) Component(I) Chart(C) Fault tree(T) Options(O) Desplay(V) Admin. data(M) Help(H)
HAZchart
HC Title Component Name Malfunction Mode HC Remarks HC Update Date FT Title Top Event FT Remarks Probability Unit Frequency
HC-0002 LCV1 Excessive Opening 2014/03/23 23:57:05 FT-0003 Gas leakage to the Liquid line 9.65E-002 /year Medium
HC-0004 LCV1 Excessive Opening 2014/03/24 00:43:39 FT-0005 Gas leakage to the Liquid line 1.16E-003 /year Low
PHA 140318test3 - [Risk Analysis Table]
File(F) Edit(E) Component(I) Chart(C) Fault tree(T) Options(O) Desplay(V) Admin. data(M) Help(H)
HAZOP
Line GuideWord Parameter Causes Consequences HC/FT Safeguards Probability Unit Likelihood
More Flow LCV1 Excessive Opening Liquid Level decreases; Gas leakage to the Liquid line HC-0002 FT-0003 LI1 9.65E-002 /year Medium
More Flow LCV1Excessive Opening Liquid Level decreases;Liquid level decreases further;Gas leakage to the Liquid line HC-0004 FT-0005 LI1,Level I/L 1.16E-003 /year Low
PHA 140318test3 - [Risk Analysis Table]
File(F) Edit(E) Component(I) Chart(C) Fault tree(T) Options(O) Desplay(V) Admin. data(M) Help(H)
What-if
Line Causes Consequences HC/FT Safeguards Probability Unit Likelihood Vf Severeness Vi
LCV1 Excessive Opening Liquid Level decreases; Gas leakage to the Liquid line HC-0002 FT-0003 LI1 9.65E-002 /year Medium 3 Medium 3
LCV1Excessive Opening Liquid Level decreases;Liquid level decreases further;Gas leakage to the Liquid line HC-0004 FT-0005 LI1,Level I/L 1.16E-003 /year Low 2 Medium 3

Outline

Business through Innovation

1. Overview of Mitsubishi Chemical Holdings
2. Motivation
3. SR activity
4. Risk evaluation method
5. Promoting discussion
6. Conclusion

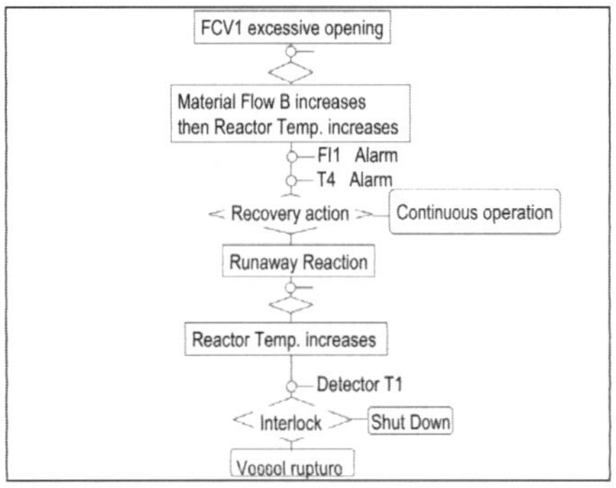

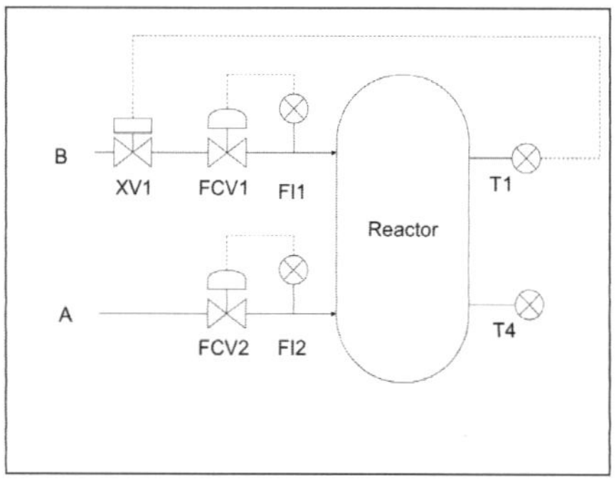

Outline

Business through Innovation

1. Overview of Mitsubishi Chemical Holdings
2. Motivation
3. SR activity
4. Risk evaluation method
5. Promoting discussion
6. Conclusion

Conclusion

Business through Innovation

- SR activity is an activity of identification of latent hazards, re-recognition of hazards and education of safety technologies to next generations.
  To facilitate SR activity, SR instructor team was formed and process safety education has been offered.

- HAZchart analysis is a tool for identification of latent hazards. It can facilitate sharing a mutual thinking process and promote discussion at a meeting.
  It is expected that these features contribute to improvement of plant staff's understanding and awareness for chemical plant safety.

Thank you for your attention!

Questions?

THE KAITEKI COMPANY

Next-Generation Growth Businesses for *KAITEKI*

20140512

These businesses contribute to Sustainability, Health, and Comfort by drawing on the Group's core technologies, and we are committed to their rapid commercialization as the drivers of our innovation strategy.

Mitsubishi Chemical Holdings

54

THE KAITEKI COMPANY

Global Network

20140512

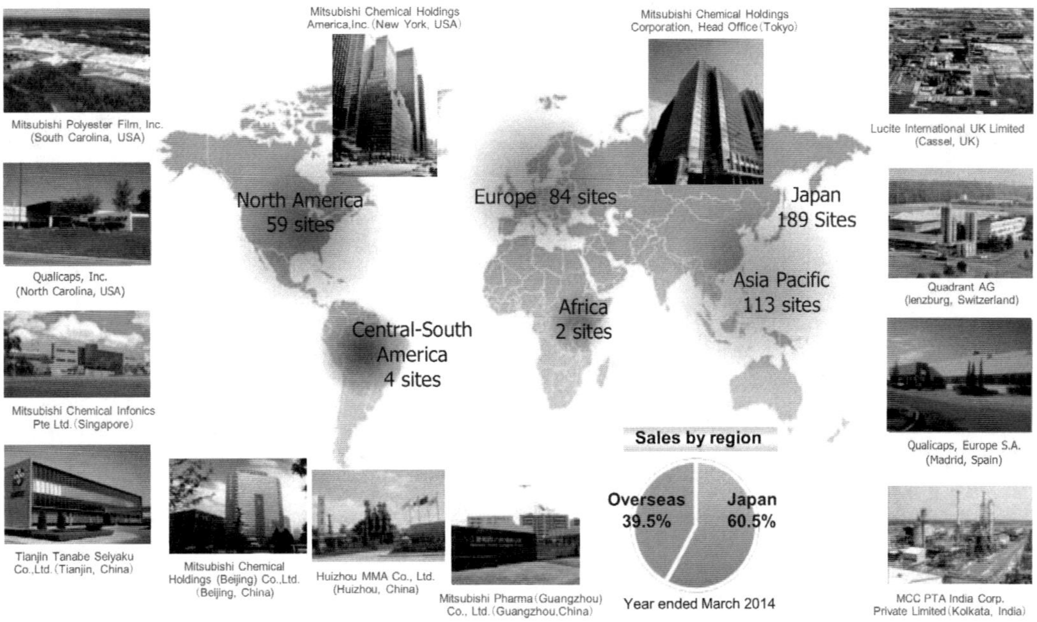

Mitsubishi Chemical Holdings

55

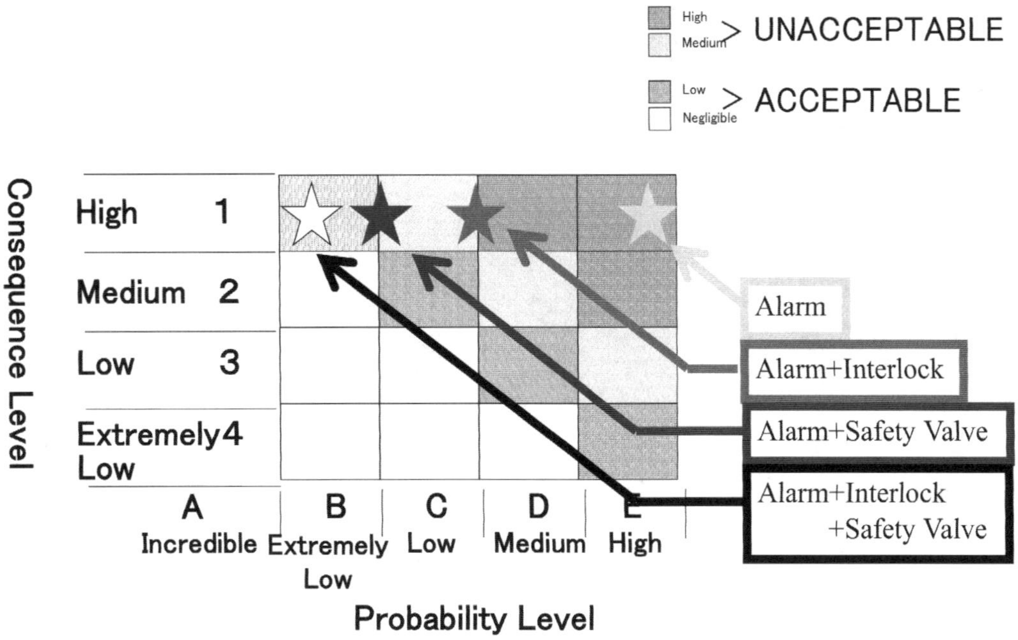

Application
-Human Error Data-

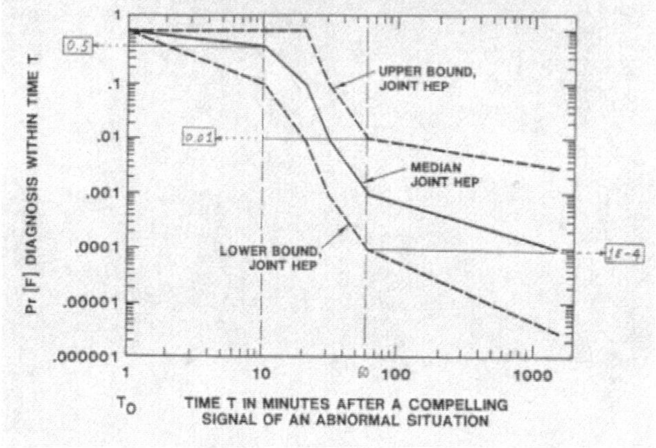

→Handbook of Human Reliability Analysis with Emphasis on Nuclear Power Plant Applications Final Report

Application
-Equipment Reliability Data-

- Process Equipment Reliability Data (CCPS)
- Equipment Reliability Database (Japan Atomic Energy Lab.)

Simple data table for screening study

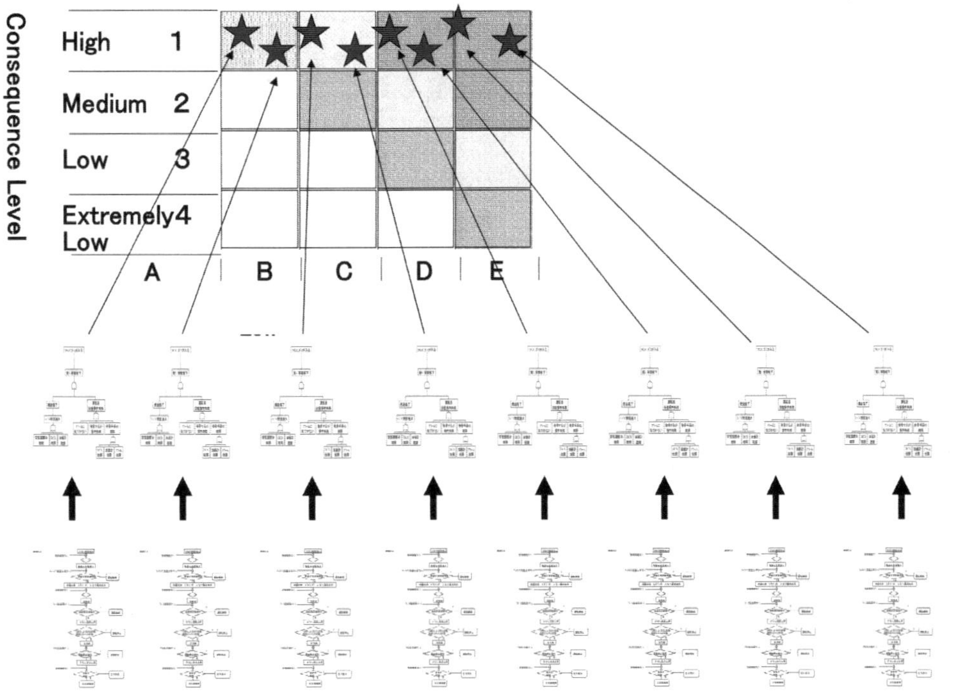

Consequence Level
High 1
Medium 2
Low 3
Extremely 4
Low
A
B
C
D
E

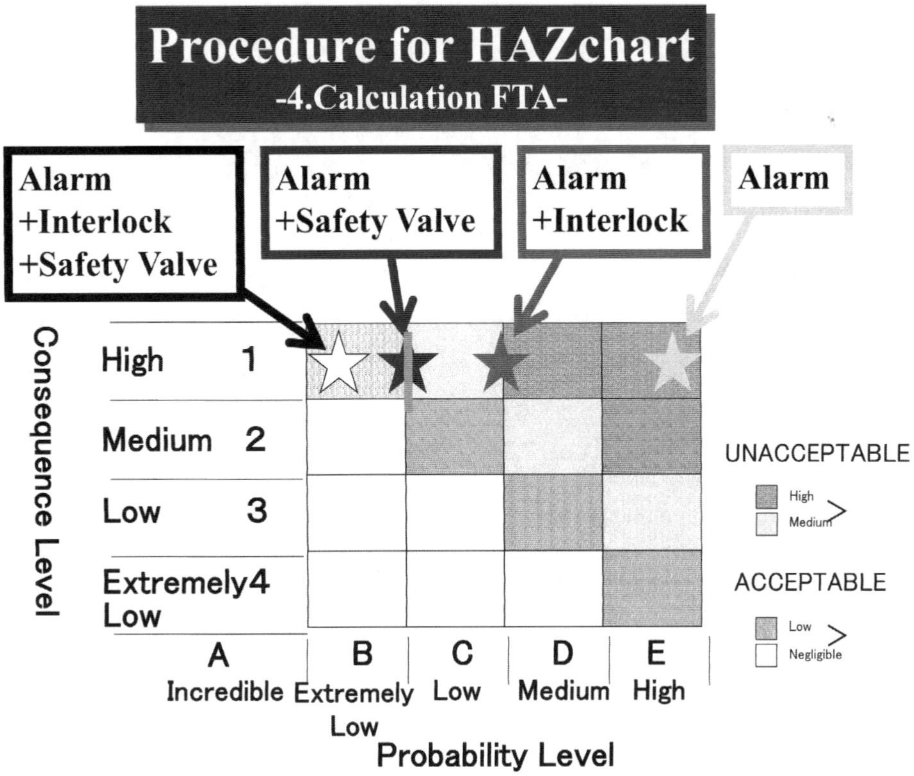

Procedure for HAZchart
-4.Calculation FTA-
Alarm
+Interlock
+Safety Valve
Alarm
+Safety Valve
Alarm
+Interlock
Alarm
Consequence Level
High 1
Medium 2
Low 3
Extremely 4
Low
A
Incredible
B
Extremely
Low
C
Low
D
Medium
E
High
Probability Level
UNACCEPTABLE
High
Medium
ACCEPTABLE
Low
Negligible

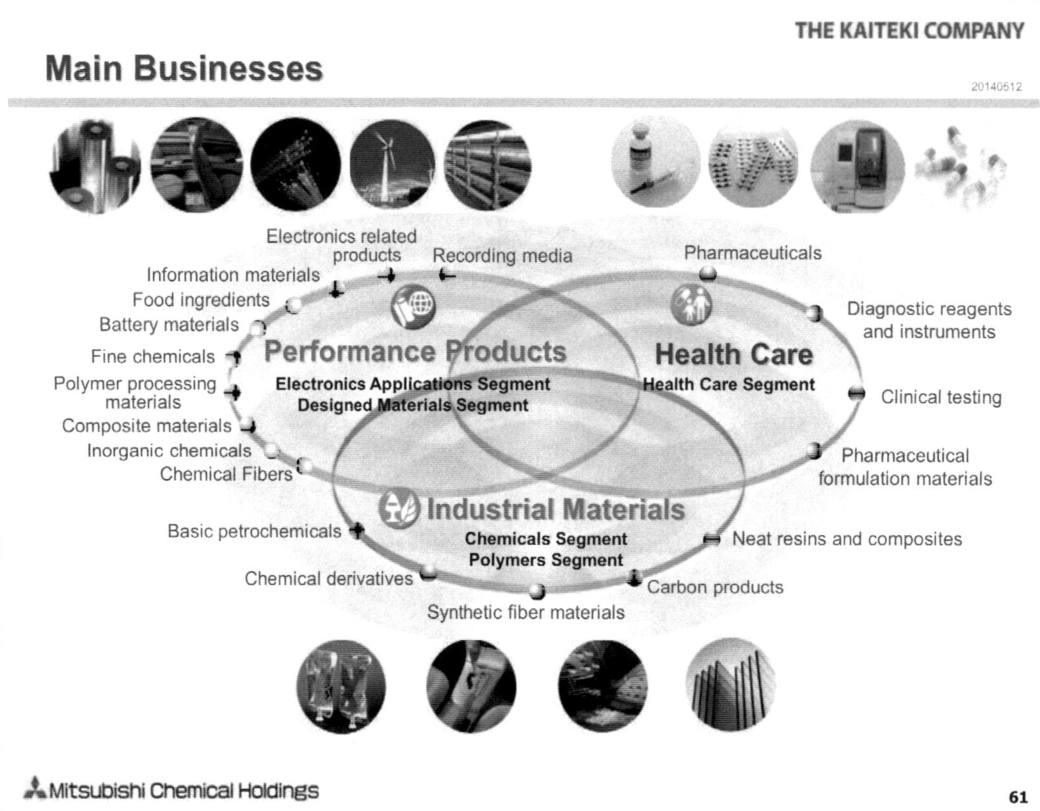

THE KAITEKI COMPANY
Main Businesses
20140512
Electronics related products
Recording media
Information materials
Food ingredients
Battery materials
Fine chemicals
Polymer processing materials
Composite materials
Inorganic chemicals
Chemical Fibers
Performance Products
Electronics Applications Segment
Designed Materials Segment
Pharmaceuticals
Diagnostic reagents and instruments
Health Care
Health Care Segment
Clinical testing
Pharmaceutical formulation materials
Industrial Materials
Chemicals Segment
Polymers Segment
Basic petrochemicals
Neat resins and composites
Chemical derivatives
Carbon products
Synthetic fiber materials
Mitsubishi Chemical Holdings
61

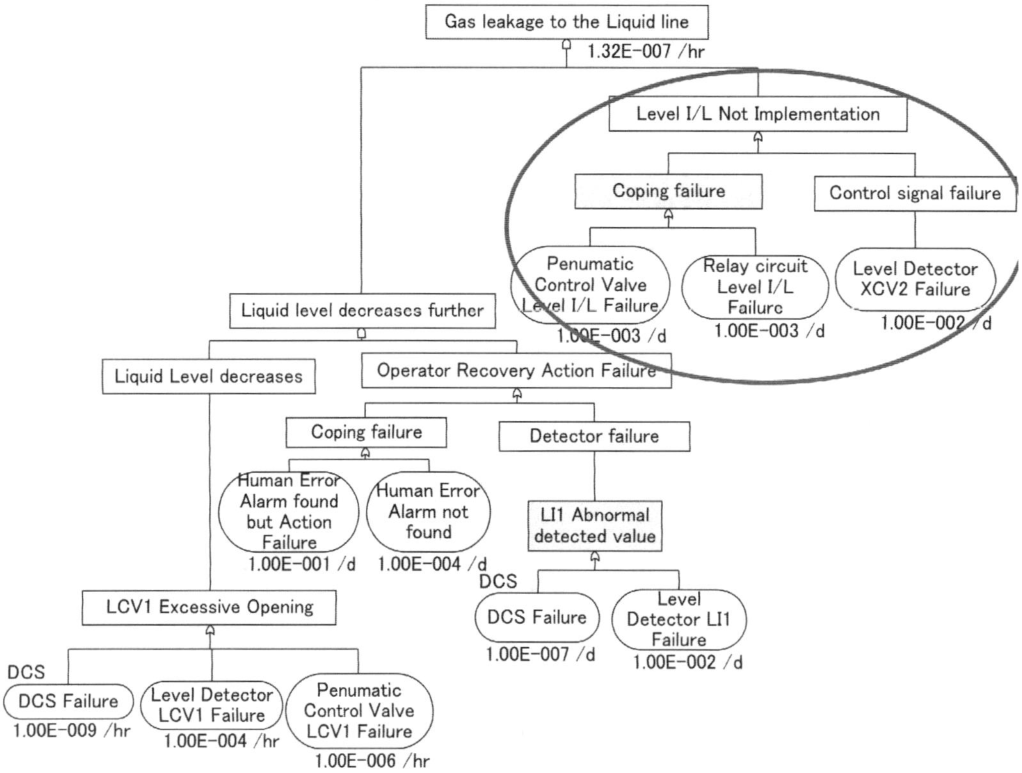

Gas leakage to the Liquid line
1.32E-007 /hr
Level I/L Not Implementation
Coping failure
Control signal failure
Penumatic Control Valve Level I/L Failure
1.00E-003 /d
Relay circuit Level I/L Failure
1.00E-003 /d
Level Detector XCV2 Failure
1.00E-002 /d
Liquid level decreases further
Liquid Level decreases
Operator Recovery Action Failure
Coping failure
Detector failure
Human Error Alarm found but Action Failure
1.00E-001 /d
Human Error Alarm not found
1.00E-004 /d
LI1 Abnormal detected value
DCS
DCS Failure
1.00E-007 /d
Level Detector LI1 Failure
1.00E-002 /d
LCV1 Excessive Opening
DCS
DCS Failure
1.00E-009 /hr
Level Detector LCV1 Failure
1.00E-004 /hr
Penumatic Control Valve LCV1 Failure
1.00E-006 /hr

2015 AIChE CCPS Asia-Pacific Conference on Process Safety

Standards and Governance to Address the Challenges of Organisational Change and Competency Assurance

Romanie Hollingworth

10 February 2015

Introduction - Legal / Regulatory Scene Setting

1. Definitions: Organisational change and competency assurance
2. Standards and governance → compliance management and corporate governance
3. When are the standards / governance principles used?
4. How do they assist here?

2

1. Definitions

- Organisational change:
  - When a company or organisation goes through a transformation; can be when business strategies or major sections of an organisation are altered
- Competency assurance:
  - Process whereby the competency of personnel is assessed and reported on
  - Provides the link between individual performance and organisational outcome

3

2. Standards and Governance

- ISO 19600-2014 Compliance Management Systems
  - Replaces AS 3806 – 2006 (dated 5 December 2014)
  - Guidance for establishing, implementing, maintaining and improving an effective and responsive compliance management system within an organisation
  - Principles-based, not prescriptive → adapt to suit entity
  - Compliance is an outcome of an organisation meeting its obligations: legal, regulatory, industry standards, ethics and community standards
  - Successful organisations have a culture of integrity and compliance → need leadership, need to embed it

4

2. Standards and Governance (cont)

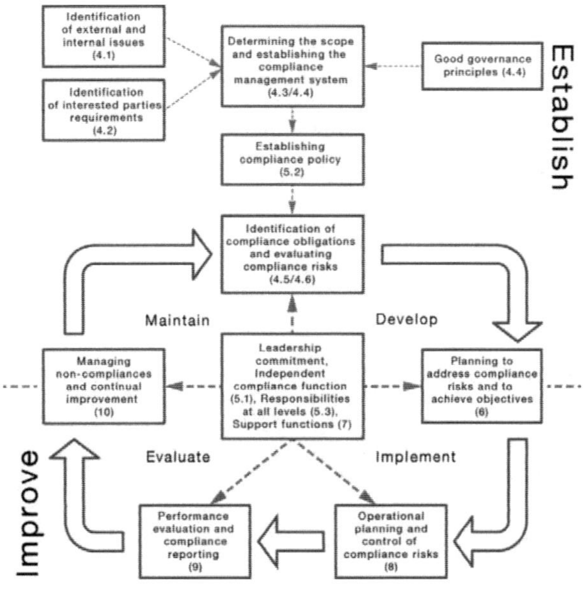

5

2. Standards and Governance (cont)

- ISO 19600-2014 is not legally binding per se:
  - Law or licence condition need to reference it for this to occur
  - Use the standard as a benchmark
- Corporate governance:
  - System by which a company / entity is directed and managed
  - Provides accountability, transparency and control systems to deal with risk
  - Increased relevance since GFC
- Board of directors – ultimate responsibility

6

3.When Are Standards / Governance Principles Used?

- ASIC Regulatory Guide 104 "Licensing: Meeting the general obligations" (AFSL) – compliance programs
- ASX Corporate Governance Principles and Recommendations – Third Edition 2014 ("if not, why not?")
- ACCC – seek court-ordered compliance program
- APRA – Prudential Standard CPS 510 Governance - binding
- Due diligence defences / defences to corporate crimes under legislation
- Compliance management systems can mitigate penalties
- ISO 19600-2014 → international standard for regulators?

7

4. How Do They Assist Here?

- Integrate governance, risk management and compliance
- System is designed to look out for change and risk and have a process to develop response strategy
- System is designed to be flexible and adaptable
- System involves self-monitoring and measuring → continual improvement
- NB organisational change can:
  - divert company's attention and resources away from compliance management system
  - itself damage the compliance management system

8

Questions

Romanie Hollingworth
Senior Corporate Law Consultant
romaniehollingworth@yahoo.com.au
0401 454 352

9

THE UNIVERSITY OF
WESTERN AUSTRALIA
FACULTY OF SCIENCE
Fitness to Operate and Safety in Changing and Unpredictable Environments
Professor Mark Griffin, Director, Centre for Safety

THE UNIVERSITY OF
WESTERN AUSTRALIA
FINANCIAL POST
Canada's oil patch bracing for coming 'retirement tsunami'
nada's oil patch is about to experience a 'retirement tsunami' — a crew change w

THE UNIVERSITY OF
WESTERN AUSTRALIA
The Sydney Morning Herald
BusinessDay
Oil slump spurs merger wave among juniors
December 22, 2014
gas sector as junior explorers struggle with the impact of declining investor support an
tched finances.

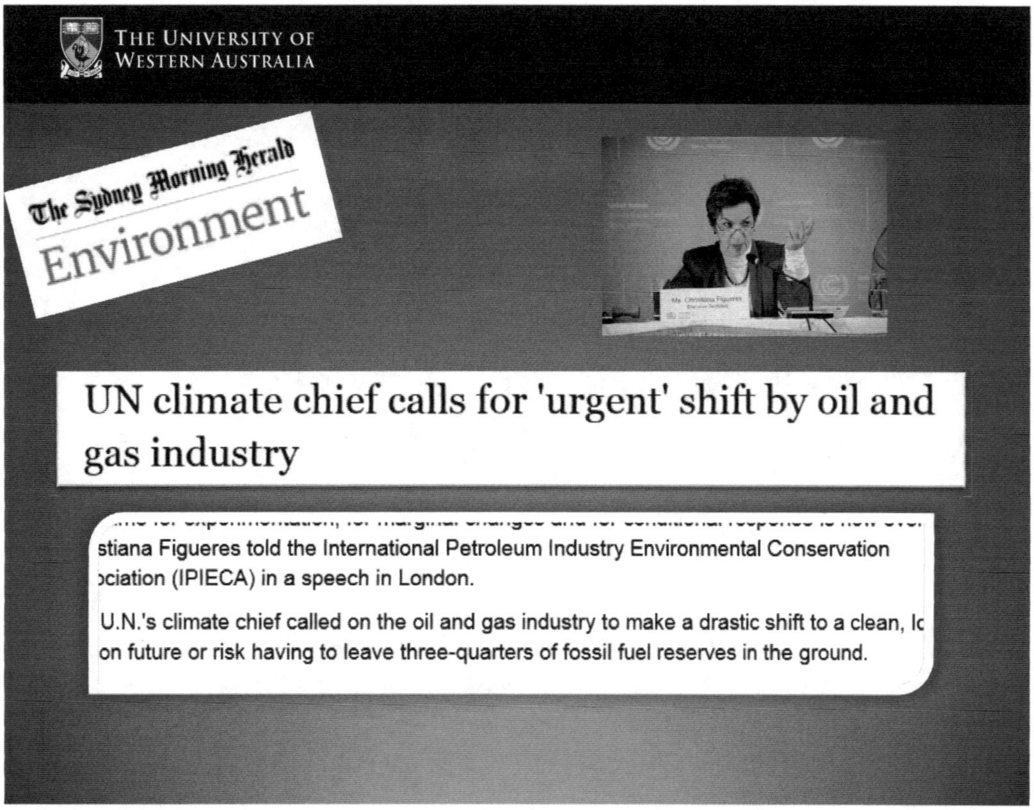
THE UNIVERSITY OF
WESTERN AUSTRALIA
The Sydney Morning Herald
Environment
UN climate chief calls for 'urgent' shift by oil and gas industry
stiana Figueres told the International Petroleum Industry Environmental Conservation
ociation (IPIECA) in a speech in London.
U.N.'s climate chief called on the oil and gas industry to make a drastic shift to a clean, lc
on future or risk having to leave three-quarters of fossil fuel reserves in the ground.

Safety in a changing world
Learn
It is vital to ask questions "What if?", "Why?", "Can you explain?", "Can you show me?", "Can you prove it?".
Hon Justice Charles Haddon-Cave, APPEA 2014
Future oriented
There is always a tendency to solve yesterday's problems
Lord Cullen, Piper25, 2013
Change
Major step change improvements that consistently use best available and safest technology are required by industry and government
Final Report Macondo Well Blowout

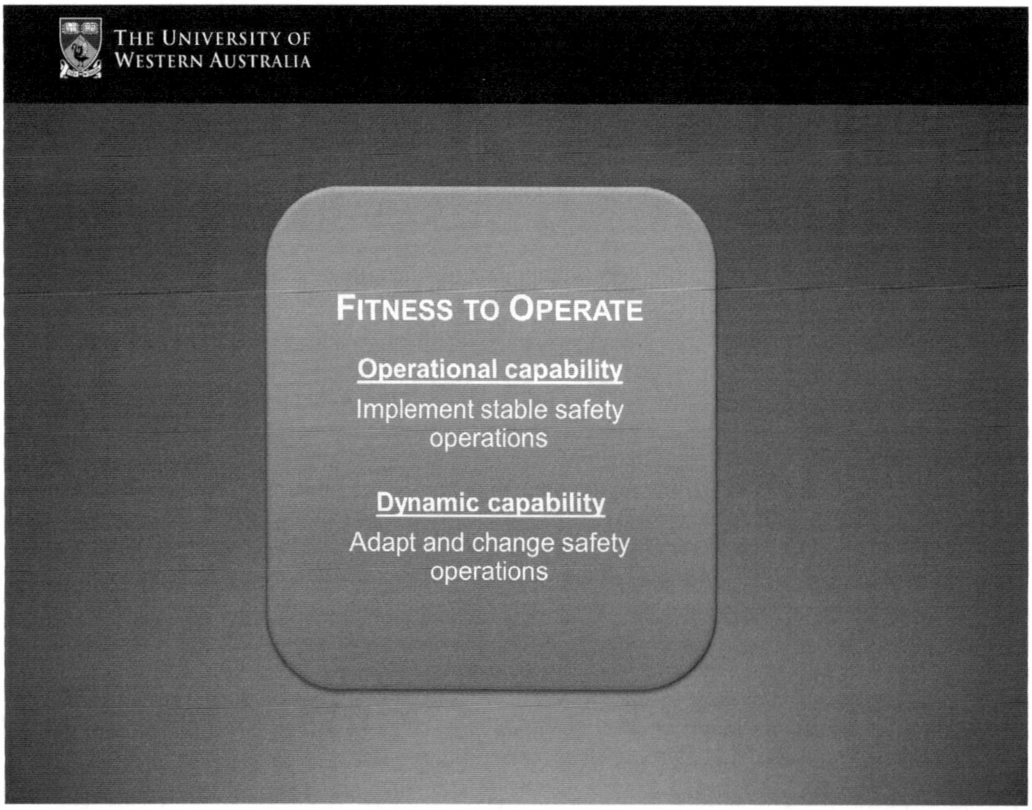

The University of Western Australia
Fitness to Operate
Operational capability
Implement stable safety operations
Dynamic capability
Adapt and change safety operations

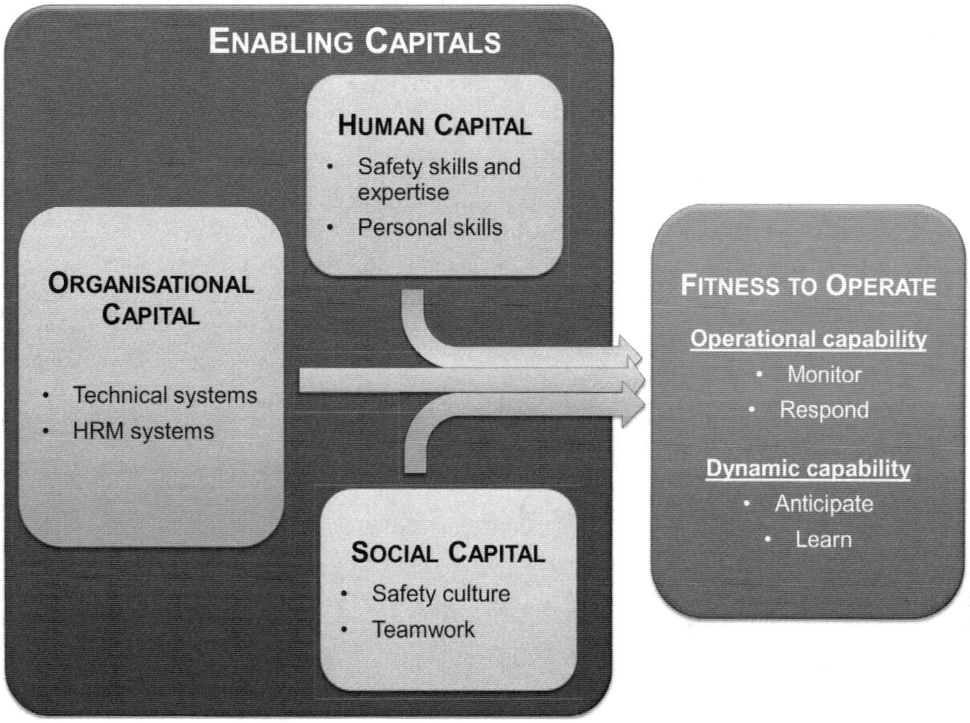
ENABLING CAPITALS
HUMAN CAPITAL
• Safety skills and expertise
• Personal skills
ORGANISATIONAL CAPITAL
• Technical systems
• HRM systems
SOCIAL CAPITAL
• Safety culture
• Teamwork
FITNESS TO OPERATE
Operational capability
• Monitor
• Respond
Dynamic capability
• Anticipate
• Learn

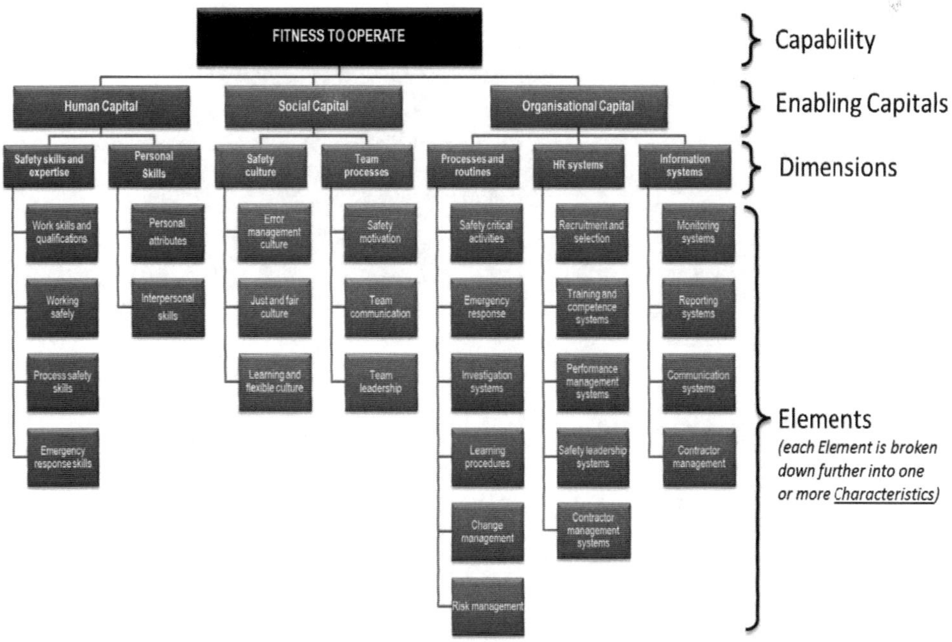
FITNESS TO OPERATE
Human Capital
Social Capital
Organisational Capital
Safety skills and expertise
Personal Skills
Safety culture
Team processes
Processes and routines
HR systems
Information systems
Work skills and qualifications
Working safely
Process safety skills
Emergency response skills
Personal attributes
Interpersonal skills
Error management culture
Just and fair culture
Learning and flexible culture
Safety motivation
Team communication
Team leadership
Safety critical activities
Emergency response
Investigation systems
Learning procedures
Change management
Risk management
Recruitment and selection
Training and competence systems
Performance management systems
Safety leadership systems
Contractor management systems
Monitoring systems
Reporting systems
Communication systems
Contractor management
Capability
Enabling Capitals
Dimensions
Elements
(each Element is broken down further into one or more Characteristics)

THE UNIVERSITY OF
WESTERN AUSTRALIA
LOST
CONFUSED
UNSURE
UNCLEAR
PERPLEXED
DISORIENTED
BEWILDERED

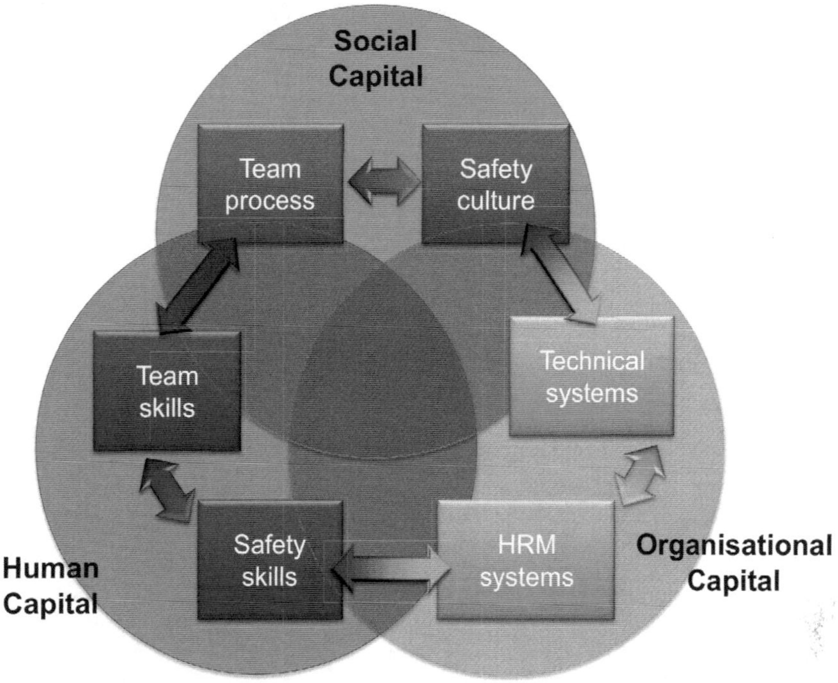

THE UNIVERSITY OF WESTERN AUSTRALIA

Dynamic safety

- Build connections
- Generate uncertainty
- Localise solutions

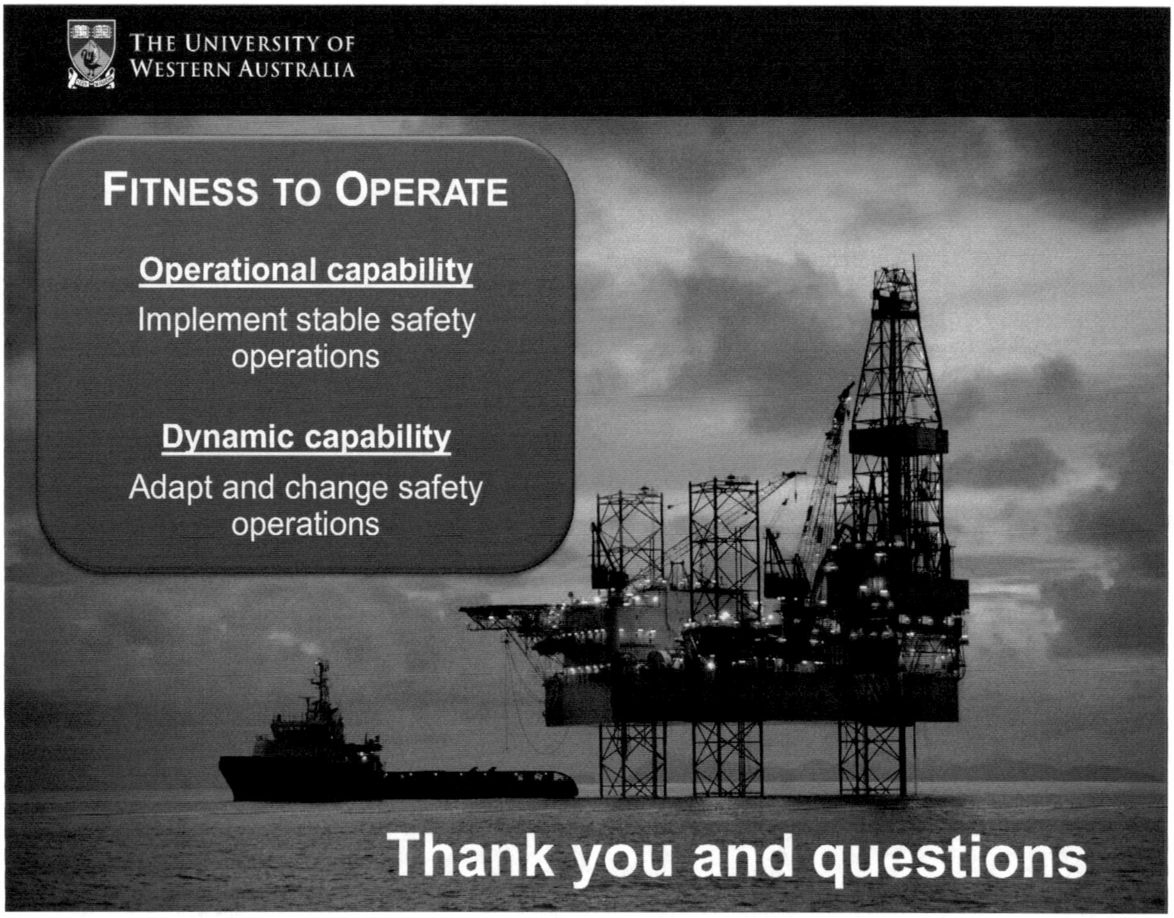
THE UNIVERSITY OF
WESTERN AUSTRALIA
FITNESS TO OPERATE
Operational capability
Implement stable safety operations
Dynamic capability
Adapt and change safety operations
Thank you and questions

Mark Small
General Manager – Offshore and Maritime

Industrial Foundation for Accident Prevention

IFAP
Safer Healthier Workplaces

IFAP
Safer Healthier Workplaces

10/03/2015 2

2010 Manila Amendments

10/03/2015 3

Navigation Act 2012

Marine Orders 70 - 73
(previously known as Marine Order 3)

- *Marine Order 70 (Seafarer certification) 2014*
- *Marine Order 71 (Masters and deck officers) 2014*
- *Marine Order 72 (Engineer officers) 2014*
- *Marine Order 73 (Ratings) 2014*

10/03/2015 4

IFAP
Safer Healthier Workplaces

30th January 2014 State of the Nation: First Quarter 2014 Slide 5

IFAP
Safer Healthier Workplaces

10/03/2015 6

IFAP

Safer Healthier Workplaces

10/03/2015 7

IFAP

Safer Healthier Workplaces

- IMCA - International Marine Contractors Association
  - Compétence Assurance and Assessment Framework
  - www.imca-int.com

10/03/2015 8

IFAP
Safer Healthier Workplaces
www.ifap.asn.au
1300 432 700
North Lake | Fremantle | Kalgoorlie | Geraldton

Make necessary changes based on lessons learned from 11.22 accident at Qingdao

Mu Shanjun, Ph.D.
vice president of SINOPEC Safety Engineering Institute
more than 30 years experience in petrochemical process safety
mushanjun@vip.sina.com

Abstract

The disaster, occurring on Nov. 22nd in Qingdao, China, resulted in 62 death knell and 136 injuring. It could be considered as a typical black swan event, revealing many management defects existing in petrochemical industry and giving us a costly warning that we should make major changes to our process safety. In this presentation, the latest investigation results about this accident will be introduced, and the introspection about the mistakes in risk assessment and process safety management will also be shared.

Contents

1. Introduction to the 11.22 accident

2. Common problems analysis

3. Major accidents should bring to us major changes

1. Introduction to the 11.22 accident

(1) Background about Dong-Huang pipeline

- Connect Dongying with Huangdao oil terminal in Qingdao.
- Laid in 1985 and put into operation in 1986.
- Design capacity is 20 million t/a and design operating pressure is 6.27 MPa.
- Total length is 248.5 km and the pipe diameter is 711 mm, made of API5LX-60 straight seam welded steel pipe.
- Covered with asphalt felt and the current cathodic protection is also utilized to avoid corrosion.
- Changed its transportation direction from Huangdao Oil terminal to Dongying with a transportation capacity of 10 million t/a in Oct., 1998.

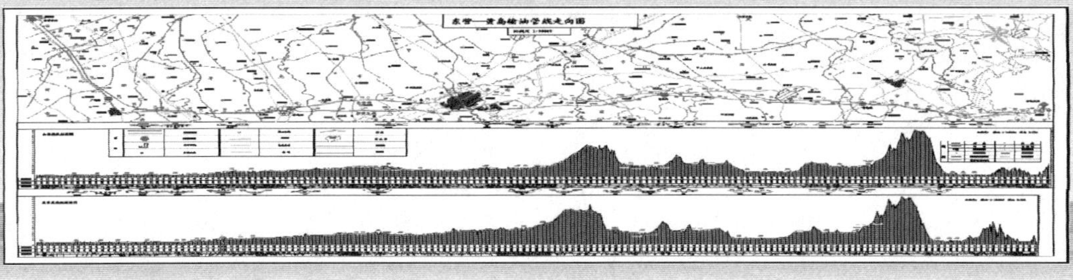

Risk control and management program of Dong-Huang Pipeline

- **1986:** put into operation.
- **2009:** Risk control and management program was carried out.
- **2010:** Rerouting application was proposed by SINOPEC to HETDZ Planning Bureau which replied orally that there was no space for rerouting, resulting in an exclusion of Huangdao segment of Dong-huang pipeline during its risk control and management program.
- **2009, 2010, 2011:** Annual inspection was carried out.

Changes of surrounding environment

- **1995:** Drainage channel was constructed by local government, open channel.
- **1997, 2002, 2008:** Drainage channel was gradually covered with concrete plates.
- **2006:** Reconstruction of Qinhuangdao Road(cross site location) was initiated.
- **2013:** Repairment of Qinhuangdao Road was commenced.

2004 2 月 2

2004 9 月 23

2007 12 月 5

2008 6 月 28

(2) Accident playback

- **Mixed crude oil:** density: 0.86t/m^3. saturated vapor pressure: 13.1kPa. explosion limits: 1.76%-8.55%. flash point in closed cup: -16°C.
- **Temperature at original terminal:** 27.8°C. Operating pressure at original terminal with full load: 4.67 MPa.

Pipeline leaking process

- **2:12 am:** pipeline pressure of Huangdao terminal's outlet declined, and was found by Weifang Control Center of SPTSC.
- **2:20 am:** Qingdao transport terminal was arranged to patrol the pipeline by Weifang Control Center.
- **2:23 am:** pipeline pressure of Huangdao terminal's outlet continued to decline. The leakage of pipeline was confirmed, and the emergency plan was initiated to cease the whole pipeline's transportation.
- **2:25 am:** ceasing process was achieved.
- **2:26 am:** linesmen rushed to the leakage spot, and call the police of Huangdao district at the same time.
- **2:50 am:** leaked oil was found around Lidong Factory.

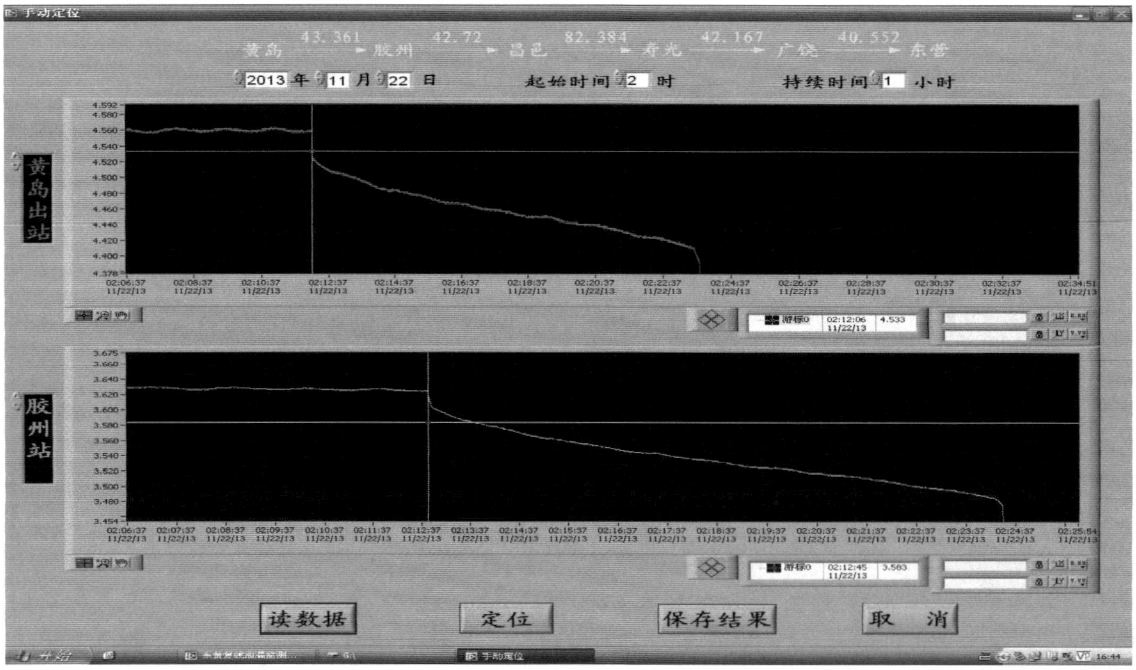

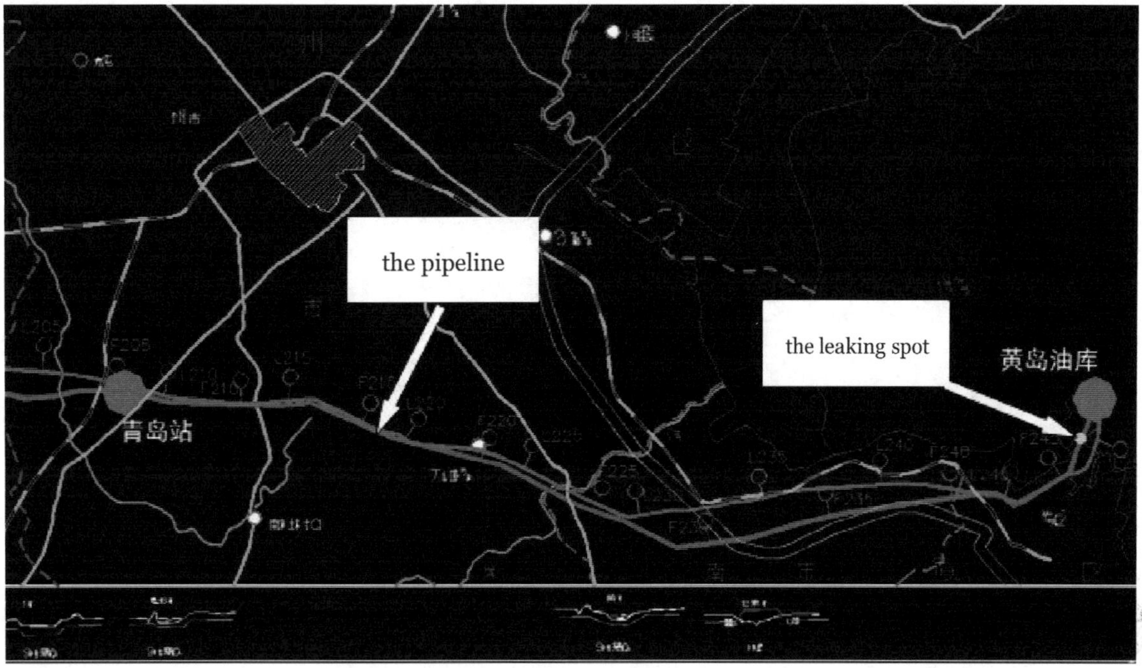

Emergency Response

- **3:40 am:** linesmen arrived at the leaking area.
- **3:45 am:** leaking spot was affirmed at the cross of Qinghuangdao Road and Zhaitangdao Road.
- **4:08 am:** spilled oil was found on the sea around the outlet of drainage pipe, the emergency plan for offshore oil spill was initiated.
- **5:00 am:** related staff from the safety, environment protection and traffic police office of Huangdao government arrived at the leaking spot early or late.
- **5:40 am:** emergency recovery crew, firefighting guardian and the operation staff arrived at the leaking spot with a total number of 30.
- **7:20 am:** a digger arrived at the spot, and began to dig around the leakage hole. The order was given by the onsite commander to open the concrete coverplate above the leakage hole. The digger followed the order to crush the coverplate with a quartering hammer.

- 10:25 am: sudden explosion happened along the 2km long drainage pipe, and the concrete coverplates were blew up. Heavy smoke was jet from the outlet of drainage pipeline on the beach followed with fire.
- The amount of leaked oil was estimated at 2000 tons.

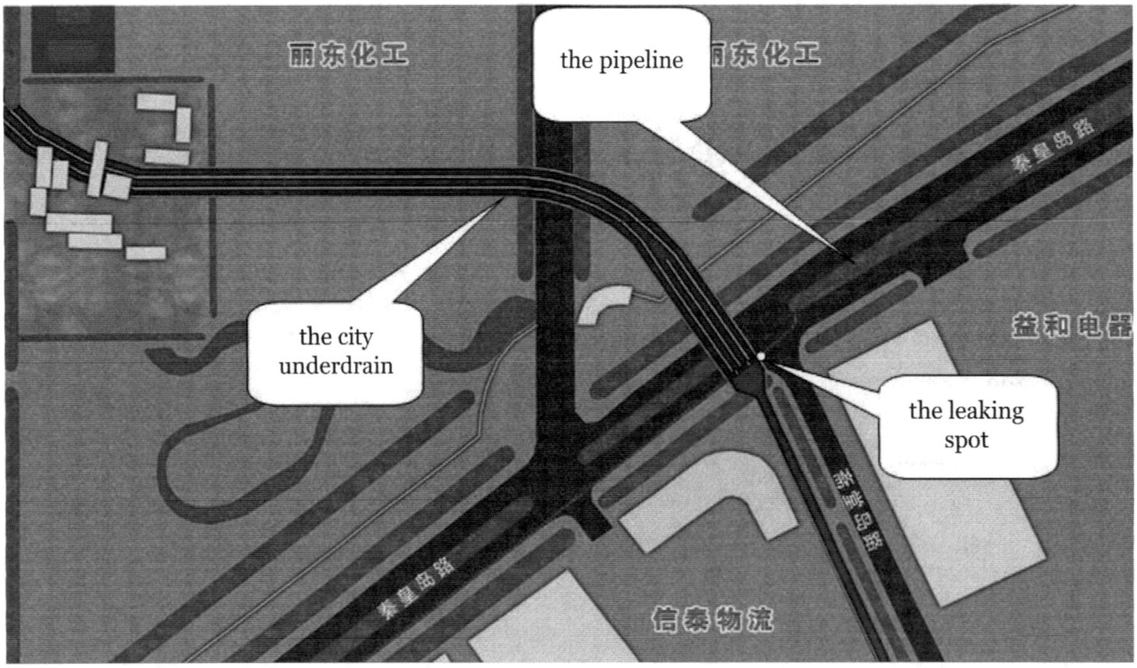

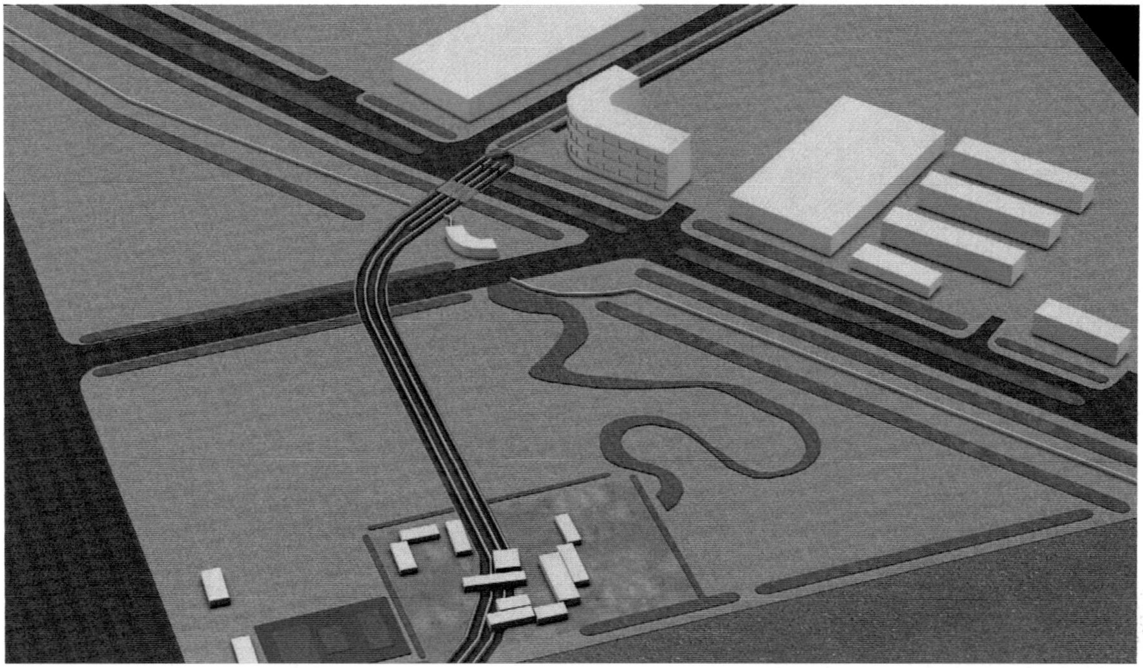

输油管线
破坏的污水渠及路面
道路
暗涵

Outlet of drainage channel on the seaside

Underdrains on the north of Lidong Chemical Co. Ltd

Destroyed container rooms.

The leaking spot

Xintai logistics office

Yihe Electrical Equipment Co., Ltd

No.2 Middle School of Huangdao

Bird view of Zhaitangdao Road

BH223E

Liugongdao Ro

An alley

HSE 监测车

(3) Cause analysis

The direct causes:

- Corrosion and thinning existed on the pipeline segment at the cross of pipeline and drainage culvert , causing the pipe burst and oil leakage.
- Crude oil flowed into the drainage culvert, overflowed onto the road.
- Digger's hammer tool was used to crush the concrete cover plate on the culvert, causing spark which ignited the explosion of vapor gas in the culvert.

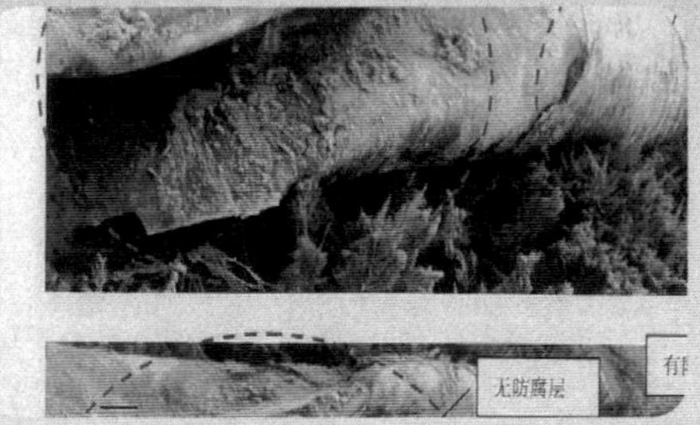

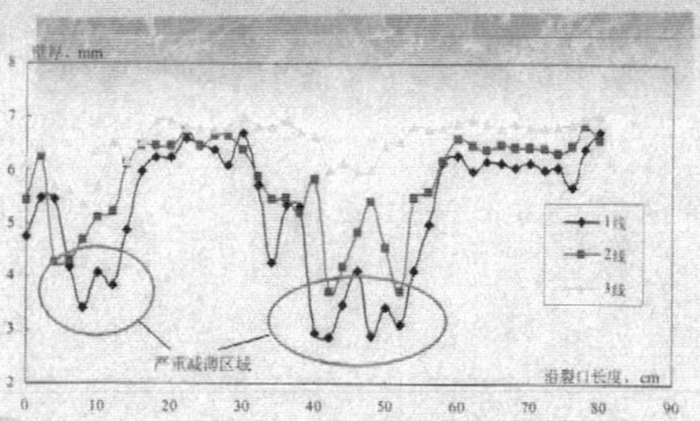

Indirect causes

- The company's safety accountability was not settled down, the risk inspection and control program was not sufficient, and the emergency response measures at the spot was not appropriate.
- The safety acts and regulations were not complied by local government.
- Pipeline protection authority has not properly taken their duty.
- The land planning by local government at the accident region was disordered.
- Risk assessment for the oil leak situation was mistaken, caused inadequate emergency response .

2. Common problems analysis

(1) Common problems of the black swan events

A black swan was a metaphor for something that could not exist. It rarely happens, but once occurred, could make big impacts. Future is unpredictable. Our world is controlled by black swan events. In order to prevent major accidents, we must try to study and understand black swan events.

Common causes behind the major accidents in history, such as Bhopal accident and bp Mexico Gulf oil release, reflect two problems, one is unaware of the risks, and another is underestimate of the risks.

Chongqing Kaixian blowout on 23rd, Dec., 2003

Bp Mexico Gulf oil release on 20th, Apr., 2010

Unaware of the risks

"11,22" pipeline accident is a typical event of such kind.

- Lack of sufficient safety data of crude oil, such as explosion limits, saturated vapor pressure.
- Unaware of underdrain channel.
- Ignoring the risks of explosion.

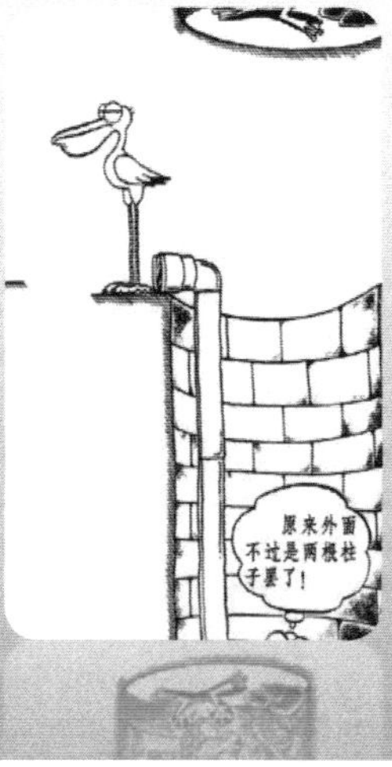

Underestimate of the risks

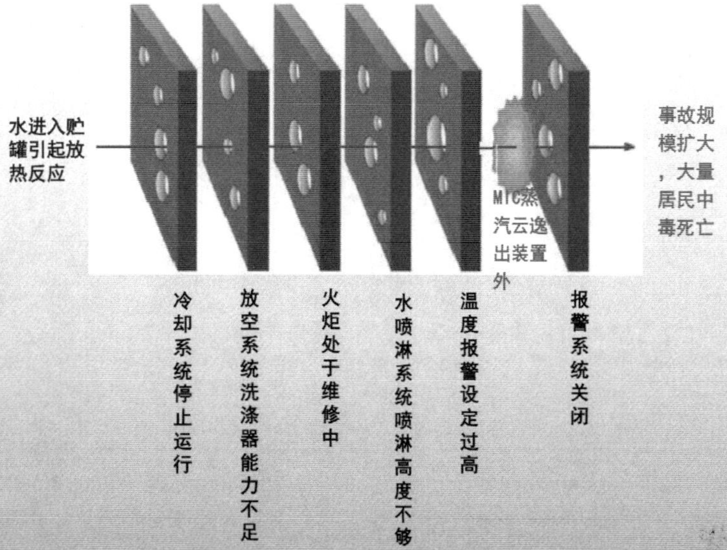

Bhopal Accident

A methyl isocyanine toxic leakage accident happened in Bhopal insecticide factory on 3rd Dec., 1984, resulting in 5000 poisoned death.

Many IPLs failed!

Bp oil release (Mexico Gulf)

At 9:47 p.m. on 20th, Apr., 2010, the biggest oil spill accident happened on deepwater horizon rig in Mexico Gulf, America, and quantities of oil spilled.

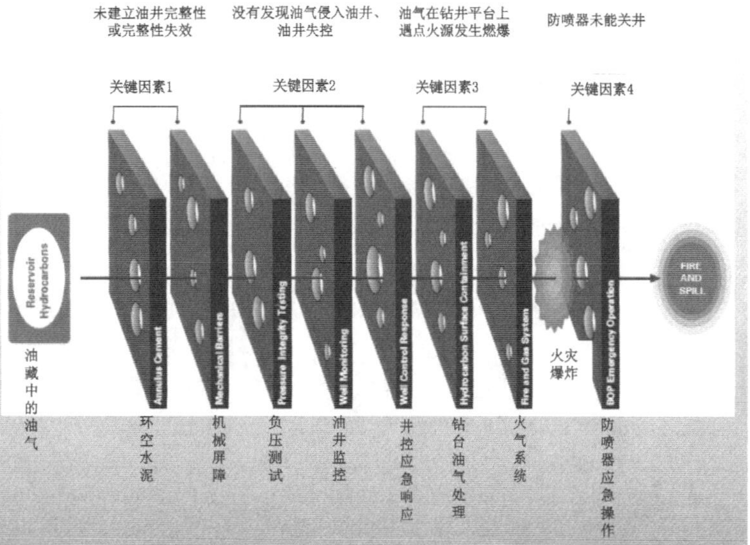

(2) Deep cause analysis

- There lies a series of casual factors in major accident. The bigger of the accident, the bigger of the number of mistakes behind the accident.
- Actually, many levels of defects exist behind a major accident, concerning with individuals, workplace, organization, regulators, and society at large.
- Risk is not fearfulness, while ignoring and underestimating risks is the biggest risk.

Defects in traditional risk assessment methods

- **Assessment based on knowledge, experience, laws and regulations is not effective.**
- **The quantitative risk assessment method based on past event frequency is unreliable.**

Defects in traditional accident analysis methods

- Focusing on the direct factors and conditions, such as unsafe acts and states.
- Improvement made just aims at settling out a single event, while comprehensive and systematic factors are not considered completely, thus similar accidents may still happen.

Defects in traditional safety management

- Static safety management is not suited to the dynamic change in risks and environment, resulting in a slow response.
- If there was no accident in the past, the management thought their measures are effective, and many bad habits gradually become acceptable and normal.
- Most of potential hazards have been found during routine safety inspection process before 11.22 accident, but no changes happened and no measures were taken, because of the inertia in management.
- The static safety management is the biggest risk for a dynamic changing risk environment.

Defects in safety culture

The defects in safety management will directly result in the decline of safety culture, producing great effects on enterprise safety belief and employee safety awareness, and then affect employee confidence, behaviors and attitude about the safety management. All these formed a vicious cycle, causing total failure of all protection layers.

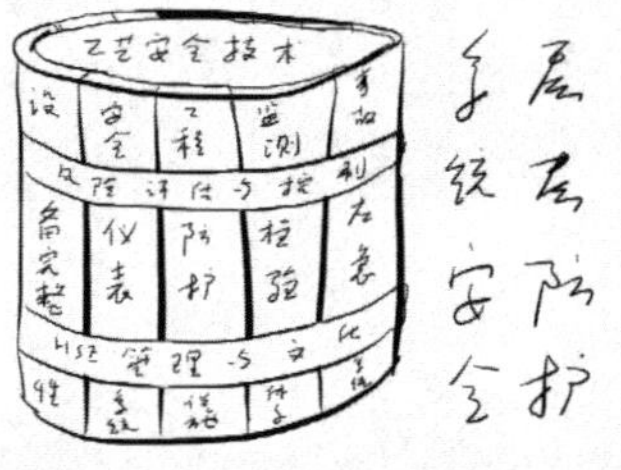

3. Major accident should bring to us major changes

The evolution of human and civilization is a press of continuous exploration. Trial and error, feedback and correction is the mechanism. Small changes were made to correct small mistakes, and social behavior rules were formulated to correct universal ones. Catastrophic events could reveal significant defects existing in our society, laws, organizations, and our behaviors, and we must make significant changes accordingly.

(1) Changes on method of risk assessment

- Generally, Risk = Consequence × Probability
- Only for occupational safety incidents,
  probability ≈ incident frequency
- Based on Murphy's Law, black swan events could and must happen some time.
- The risks of black swan events were easily and often underestimated, and corresponding prevention measures could not meet the actual requirements. That puts everyone in big danger, and we should not prevent black swan events just by good luck.

(2) Improve accident investigation methods

- Traditional accident investigation in China: focusing on the analysis of direct causes, insufficient consideration of the overall, systematic fundamental factors. The investigation is aim to find out whom to blame, limited to the accident scene basically, and conclusions are obtained in very short time.
- Suggestion: referencing the American CSB practices, carry out systematical research on overall factors relating to an accident, such as leakage, fire and explosion, emergency response, etc., and making professional suggestions on changes of policy, regulation, supervision, management, operations, behaviors and so on.

(3) Make scientific city development planning

- Chemical industry park must be effectively separated from densely populated area.
- The layout of underdrains and oil/gas pipelines in city must be scientifically demonstrated, and effective safety measures are necessary.
- Government work safety supervision methods need to be changed considering the market mechanism.

(4) Enhance the safety culture

Successful experiences and cases

- broken window theory
- subway security situation improved rapidly in New York

Changes on management

- positive safety attitude
- zero tolerance attitude
- HSE observation
- system management
- dynamic management
- continuous improvement
- specialized safety inspections regularly

(5) Improvement on emergency response

- Making enterprise-government linkage emergency plan.
- Improving the coordination and effectiveness of the emergency drills.
- Command, communications and rescue on scene need improvement, considering the significantly rising confusion with the expansion of the accident.
- Making special emergency plans for black swan events.

Q&As?

Thank You!

Financial institutions
Energy
Infrastructure, mining and commodities
Transport
Technology and innovation
Life sciences and healthcare

NORTON ROSE FULBRIGHT

Are Legal Processes Compatible with Improving Process Safety?

Michael Tooma
Partner – Head of Occupational Health, Safety and Security (Asia Pacific)

2015 AIChE CCPS Asia-Pacific Conference - 11 February 2015

Presentation Outline

- **Understanding Process Safety**
- **Fear of legal liability**
- **What is legal professional privilege?**
- **Role of legal professional privilege**
- **Is the document privileged?**
- **Waiver of privilege**

2

Montara Wellhead Platform Oil Spill
• The 21 August 2009 oil spill from the Montara Wellhead Platform was the worst spill in Australia in over 20 years
• It was the worst oil spill of its kind in the Australian offshore petrol industry
• For a period of just over 10 weeks, oil and gas continued to flow unabated into the Timor Sea

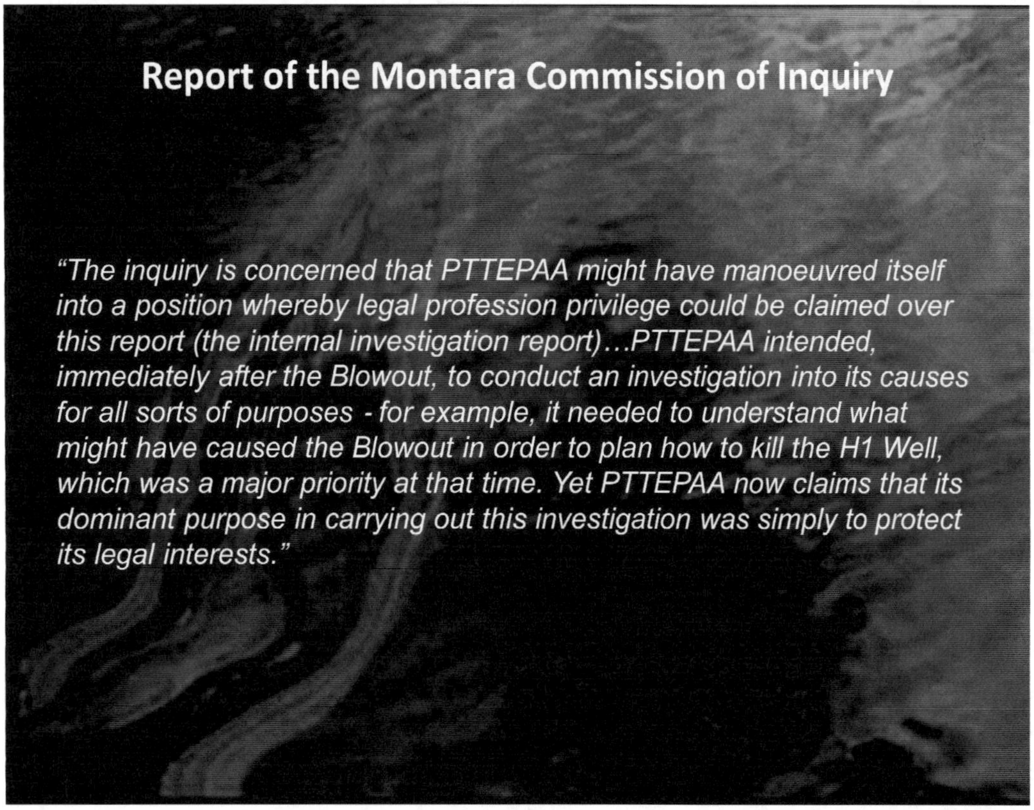
Report of the Montara Commission of Inquiry
"The inquiry is concerned that PTTEPAA might have manoeuvred itself into a position whereby legal profession privilege could be claimed over this report (the internal investigation report)...PTTEPAA intended, immediately after the Blowout, to conduct an investigation into its causes for all sorts of purposes - for example, it needed to understand what might have caused the Blowout in order to plan how to kill the H1 Well, which was a major priority at that time. Yet PTTEPAA now claims that its dominant purpose in carrying out this investigation was simply to protect its legal interests."

Understanding Process Safety

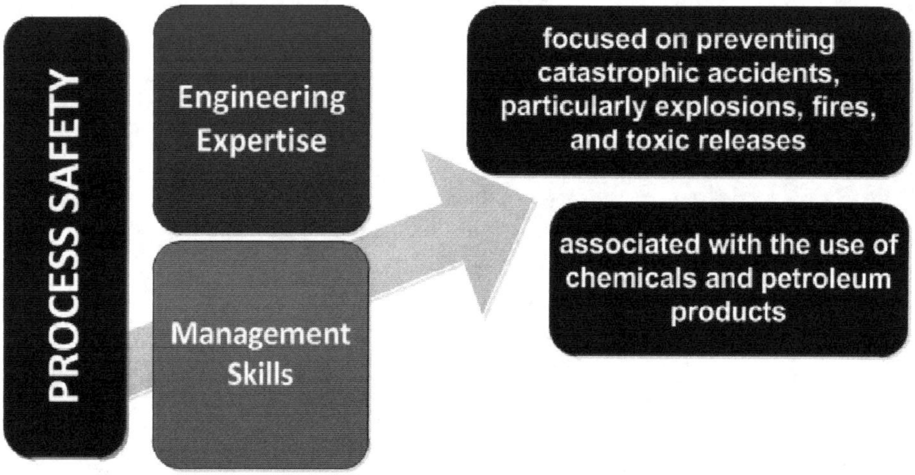

5

Fear of legal liability

"the sheer threat of judicial involvement is enough to make people think twice about coming forward with information about an incident they were involved in"

Dekker, S 2007, *Just Culture: Balancing Safety and Accountability*

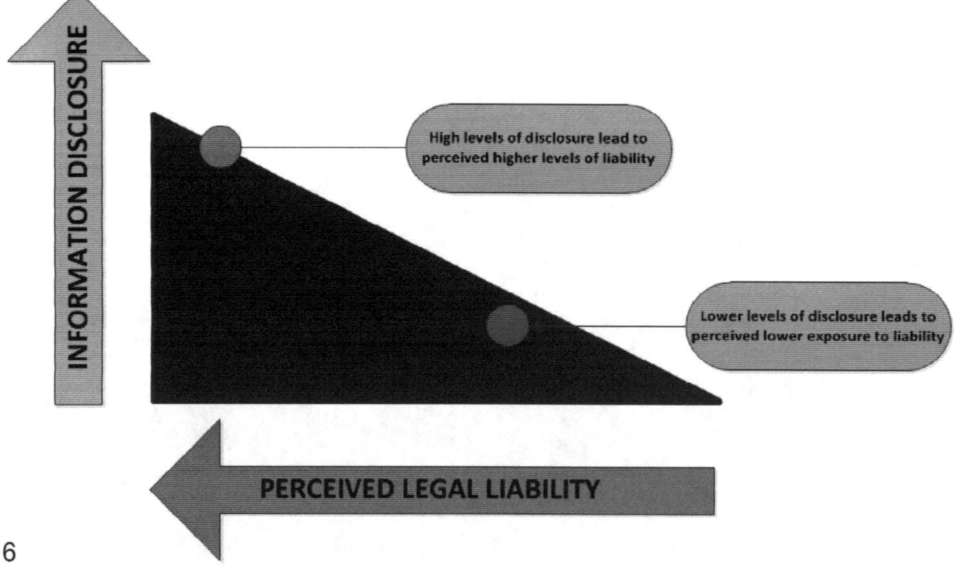

6

What is legal professional privilege?

- Legal professional privilege protects the disclosure of certain communications between a lawyer and a client when these communications are for the dominant purpose of seeking or providing legal advice, or for use in existing or anticipated legal proceedings

- In most cases legal professional privilege will be claimed over documents, but it also applies to oral or electronic communications between a lawyer and their client

- Can include those between a client and a third-party e.g. consultants if the dominant purpose test is met

7

Role of legal professional provision in addressing fear of liability

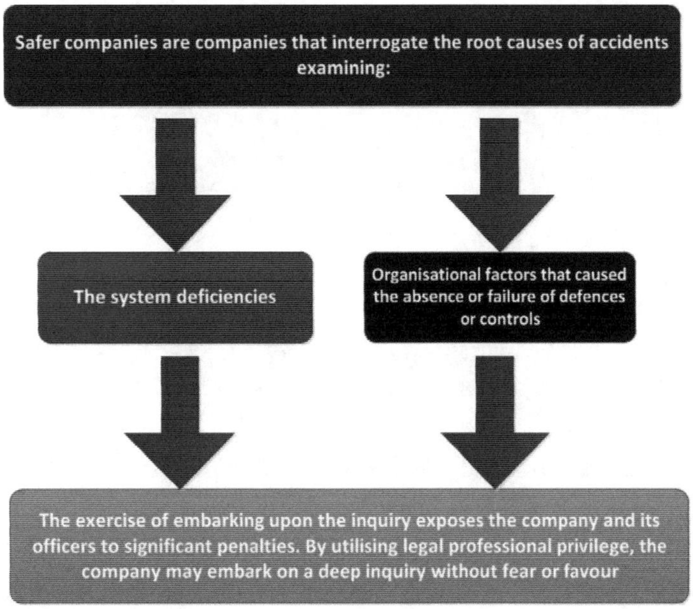

8

Elements of Proper Incident Investigation

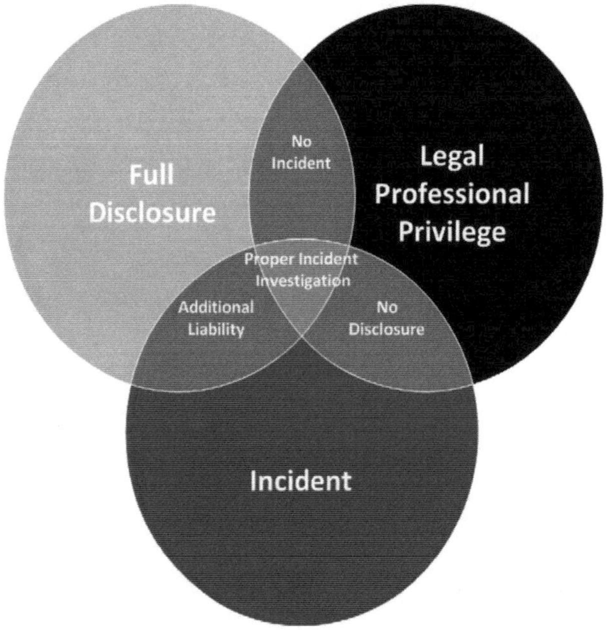

9

Waiver of Legal Professional Privilege

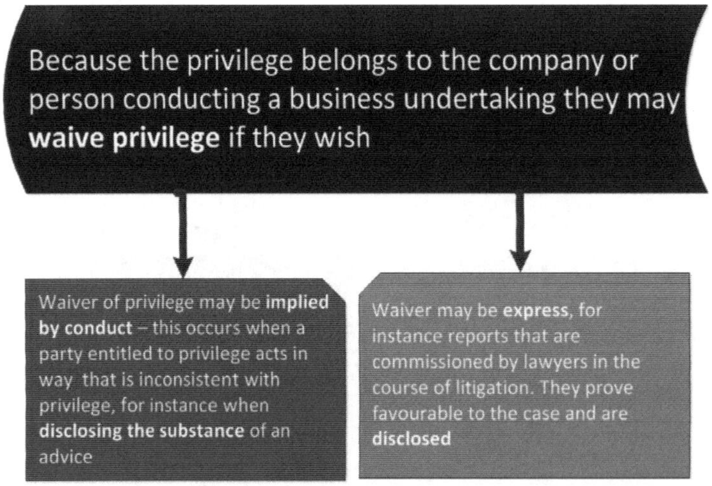

1

Is the document privileged?

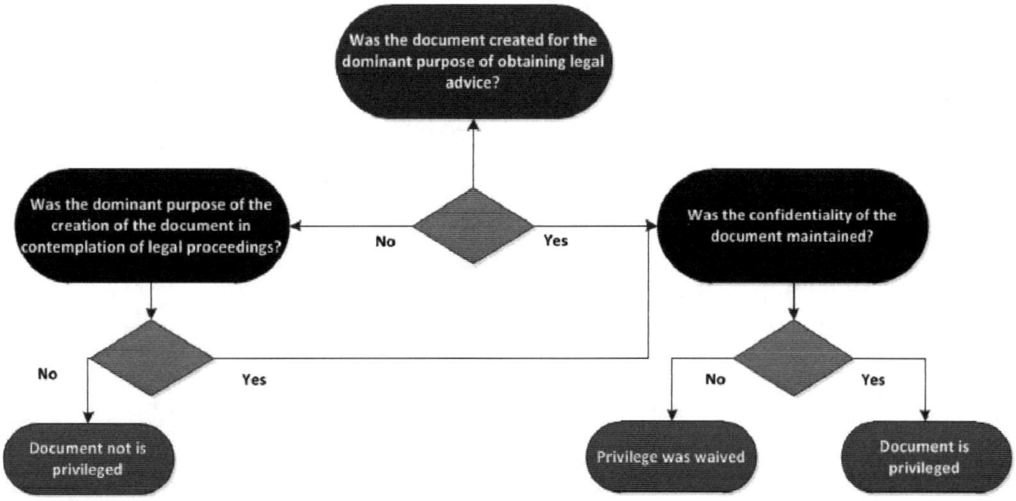

11

Questions and Discussion

12

International

*associate office

13

Disclaimer

Norton Rose Fulbright LLP, Norton Rose Fulbright Australia, Norton Rose Fulbright Canada LLP, Norton Rose Fulbright South Africa (incorporated as Deneys Reitz Inc) and Fulbright & Jaworski L.L.P., each of which is a separate legal entity, are members ("the Norton Rose Fulbright members") of Norton Rose Fulbright Verein, a Swiss Verein. Norton Rose Fulbright Verein helps coordinate the activities of the Norton Rose Fulbright members but does not itself provide legal services to clients.

References to "Norton Rose Fulbright", "the law firm", and "legal practice" are to one or more of the Norton Rose Fulbright members or to one of their respective affiliates (together "Norton Rose Fulbright entity/entities"). No individual who is a member, partner, shareholder, director, employee or consultant of, in or to any Norton Rose Fulbright entity (whether or not such individual is described as a "partner") accepts or assumes responsibility, or has any liability, to any person in respect of this presentation. Any reference to a partner or director is to a member, employee or consultant with equivalent standing and qualifications of the relevant Norton Rose Fulbright entity.

**The purpose of this presentation is to provide information as to developments in the law. It does not contain a full analysis of the law nor does it constitute an opinion of any Norton Rose Fulbright entity on the points of law discussed.
You must take specific legal advice on any particular matter which concerns you. If you require any advice or further information, please speak to your usual contact at Norton Rose Fulbright.**

14

The Community Right to Operate

AIChE CCPS 2015 Asia-Pacific Conference
(9- 11 Feb'15 , Parmelia Hilton Perth , Western Australia)

Mr. Worachai Puvisitkul
SHE Operation manager
SCG Chemical Co.,Ltd.

SCG

Agenda

- Company overview
- Current incidents and regulatory enforcement in Thailand
- Enforcement of environmental regulation
- Our PSM model and approach
- High level PSM Implementation
- PSM outcome
- Key challenge
- Summary

SCG Chemicals Overview

- SCG Chemicals Co., Ltd. ("SCG Chemicals") is the holding company for SCG's chemicals businesses established in 1986.
- One of ASEAN's largest integrated producers of upstream and downstream petrochemical products, with 2 world-scale crackers located in Thailand and 30% stake in the only cracker in Indonesia
- Successful long term partnership with world-class leaders such as Dow Chemical-USA, Mitsui Chemicals-Japan, and Mitsubishi Rayon-Japan.

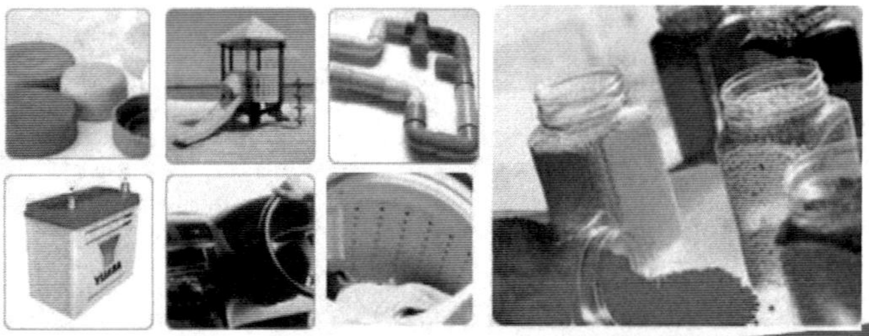

CG 2015 3 SC

Our Business Chain

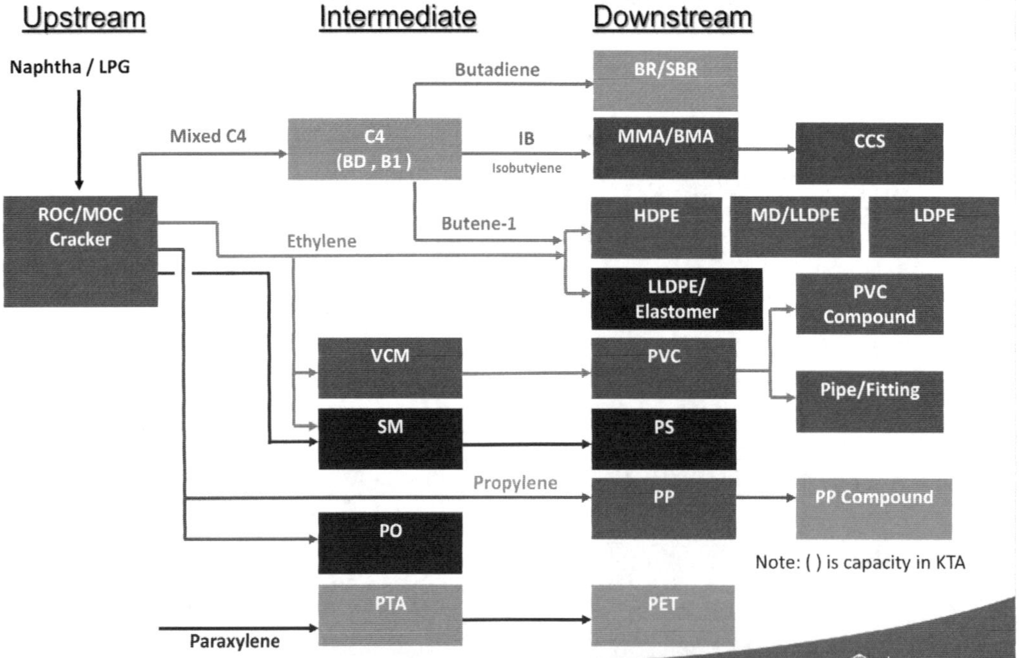

@ SCG 2015 4 SCG

SCG Chemicals Group
...over 60 companies & more than 5,000 employees

Subsidiaries

CRACKER
- Rayong Olefins
- Map Ta Phut Olefins

PE/PP
- SCG Plastics
- SCG Performance
- Thai Polyethylene
- SCG ICO Polymers

PVC
- Thai Plastics and Chemicals
- Nawa Plastics

SERVICE
- Map Ta Phut Tank Terminal
- REPCO

SCG-Dow JV

LLDPE
- Siam Polyethylene

SM/PS
- Siam Styrene Monomer
- Siam Polystyrene

ELASTOMER
Siam Synthetic Latex

PO
- MTP HPPO Manufacturing

SERVICE
- Rayong Tank Terminal

SCG-Mitsui JV

PP Compound
- Grand Siam Composites
- Mitsui Advance Composite (Zhongzan)

Film
- Siam Tohcello

PTA/PET
- Siam Mitsui PTA
- Thai PET Resin

SCG-MRC JV

MMA/CCS
- Thai MMA

Others JV

C4 & BR/SBR
- Bangkok Synthetics

MELAMINE
- Thai MFC

Overseas

INDONESIA
- (Complex) Chandra Asri Petrochemical
- (PVC) PT TPC Indo Plastic and Chemicals
- (Terminal) PT Siam Maspion Terminal

VIETNAM
- (PVC) TPC Vina
- (Compound) Viet Thai Plaschem, Chemtech

USA
- (Technology) GTC

CG 2015 5 SC

Community around Petrochemical Industrial

- Established in 1989
- No. of Manufacturing 64
- Community nearby 52
- Population 130,000 Persons

Industrial estate area.

Community.

SCG CHEMICALS SCG Operation site .

@ SCG 2015 6 SCG

Some Major Incidents in Maptaput Industrial Estate, TH

- 2008 Cumene chemical release from piping, 112 injured
- ***2009 Butene-1 leak at safety valve on ship unloading, 118 injured***
- 2010 NaOCl release from storage tank, 143 injured
- ***2012 Butadiene rubber plant fire and explosion, 12 fatal 130 injured***
- 2012 Chlorine toxic gas release , 133 injured and 500 evacuation
- 2013 Crude oil spill to sea from ship to shore tank, effect to ecology , fishing community, tourism business
- ***2014 MMA plant chemical release and fire, no injured***
- 2014 Refinery plant fire , no injured

The incident still continue to occur....

More stringent on environmental regulation are enforced

2012-2014

- Volatile organic compounds (VOCs) Control limit standard
- New specific emission of 1,3 Butadiene, Benzene standard
- Underground water contamination regulatory control
- EHIA regulation for new facilities which require more involvement of community.
- PRTR - Pollution Release transfer Register
- Annual SHE audit by government team include representative from community
- (Drafting) Plant integrity audit protocol by government and 3rd party base on PSM OSHA requirement

Our Process safety Journey to go beyond regulatory compliance

Our Objective :
Minimize risk to avoid catastrophic event thru safe operation

Our Key success :

- Build a safety culture.
- Management leadership and commitment
- Implement integrated PSM
- Implement Operational Discipline (OD)

Our PSM Model with 14 elements

Technology
- PSI
- PHA
- Safe work practice/SOP
- Management of change

Facility
- Quality assurance
- PSSR
- Mechanical Integrity
- Management of change

Personel
- Training
- Contractor
- Incident Investigation
- Management of change
- Emergency response
- Auditing

Management Leadership and Commitment

Implementing process safety management

1. Established safety culture
2. Established safety leadership demonstration through policy, principle, auditing etc.
3. Organization change
   - Leadership team at corporate level
   - Implementation team at company level
   - PSM experts group
4. Clearly identify employee safety KPIs and Process safety KPIs
5. Develop corporate PSM standards base on world class practices
6. Train to related employee including contractors
7. Implement such standards to field work
8. Track performance both leading and lagging indicators
9. Conduct 1st and 2nd party auditing
10. Continuous improvement
11. Implement operational discipline

CG 2015 11 SC

Implementing process safety metrics

Lagging :

- Fire & Explosion
- LOPC
- Property damage
- Environmental
- Compliance

Leading :

- Management audit
- Near-miss reporting
- Unsafe act/Unsafe condition reporting
- Closed recommendation of each incident on time
- Etc.

Management Field audit for Multiple points of failure

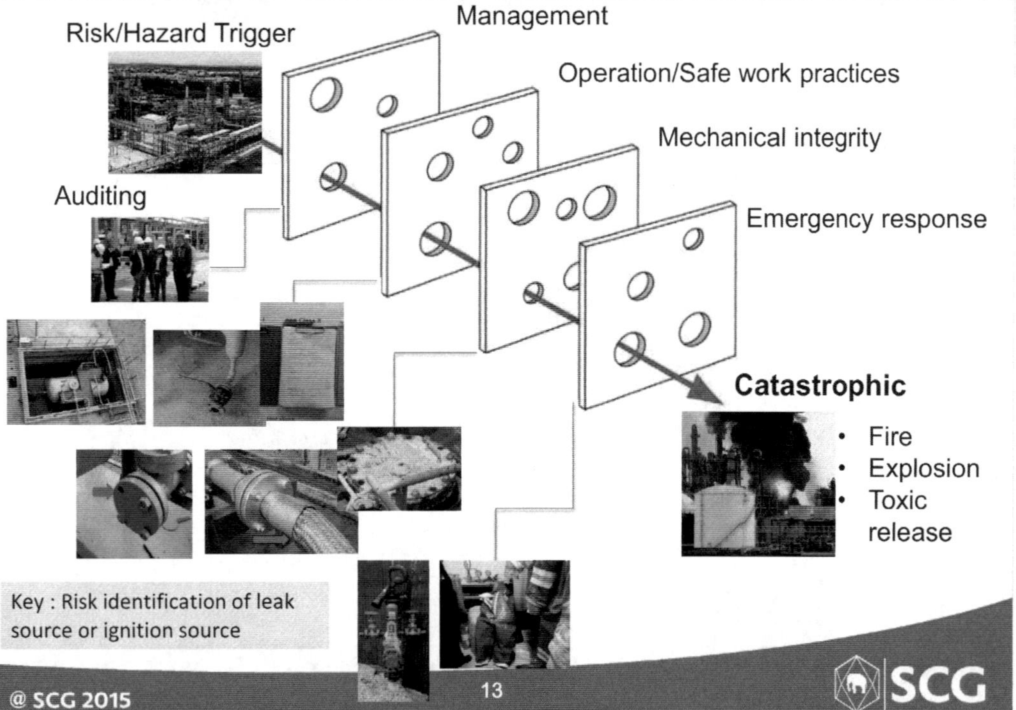

Key : Risk identification of leak source or ignition source

@ SCG 2015 13 SCG

Key challenges

- Mindset and behavior change i.e. priority production/cost and safety , reactive to proactive.
- Need effort of management involvement and consistency
- Encourage employee participation in positive
- Building competency on PSM knowledge and skill
- Need support from Top Management on resources i.e. capital , time , manpower , IT
- Sustain effective PSM in active

Key summary

- We believe that power of PSM will make trust and confident for community.
- PSM is one of our business core value to protect employee, community and business asset.
- Integration of PSM into operation can achieve business performance excellence

CG 2015 15 SC

Thank you for your attention

@ SCG 2015 16 SCG

appea

Safety Cases & Making Process Safety Personal

Miranda Taylor
APPEA Director Safety, Environment and Operational Performance
(Contact: mtaylor@appea.com.au)
11 February 2015

Making Process Safety Personal

- Process safety has seen much focus since the BP Texas City incident in 2005
- AIChemE has done much good work to progress process safety through CCPS
- This includes the development of tiered metrics that are being adopted globally
- Australia is also active in progressing the process safety agenda

appea Making Process Safety Personal

Major Challenges:

- Adopt rather than duplicate technical excellence
- Make that excellence accessible to all stakeholders by busting jargon and challenging views of risk
- Deliver simplicity from complexity
- Build an industry culture that facilitates operational discipline (interventions are taken when needed)
- Collaborate to ensure sustainable and efficient operations in a cost-constrained world

appea Making Process Safety Personal

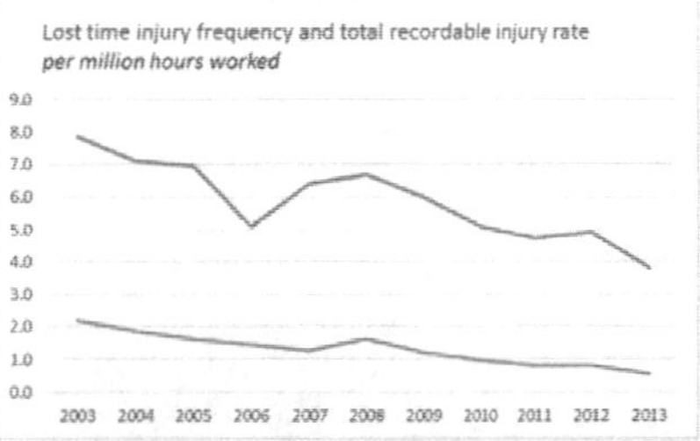

Australian Context:

- Legislative approach gives flexibility
- Increasingly complex safety cases diminishing value
- Huge growth driven by LNG
- Historical events - Longford, Varanus Island, Montara & Macondo
- Australian industry's safety journey
- IChemE Process Safety Centre
- Stand Together For Safety

Stand Together For Safety

stand together for safety
an australian oil and gas industry initiative

- *The objective of STFS is the "prevention of major accident events" through the collaboration of operators, contractors, industry groups, frontline personnel, technical experts and regulators*

STFS Steering Committee
- Provide stewardship of delivery against objectives
- Agree resource requirements
- Rank priorities and agree timeline
- Monitor progress against timeline
- Membership-Operators, Industry Groups, Contractors, Tech Experts, Frontline & Regulators

Working Groups
- Create Tools
- Integrate expertise eg IChemE, HSO
- Forum for technical discussions
- Membership- Subject Experts from Industry and Regulators

Workshops & Continuous Improvement Forums
- Facilitated workshops with cross section of Industry and Regulators
- Identify Improvement Opportunities
- Agree priorities

Collaboration

stand together for safety

STFS	Safer Together	HSO Committee	IChemE
An initiative to drive safety improvements by engaging operators, contractors, workforce and regulators	An initiative to drive safety improvements facing the CSG industry	A representative group of industry experts seeking continuous improvement	A multi-national centre of expertise building academic knowledge and tools
First priority is Process Safety Future phases may address other areas, e.g. Drilling	Focused on specific CSG issues, including personal safety	Linked to special groups, e.g. Process Safety Human Factors Aviation Well Integrity	Linked to international centres of excellence A natural home for technical info and tools, e.g. process safety animations

APPEA provide the link to foster collaboration & avoid duplication

appea Making Process Safety Personal

- Technical excellence abounds
- An interconnected world facilitates effective collaboration
- The growth of the Australian oil and gas industry provides an opportunity to provide process safety leadership
- We look forward to exporting best practice in making process safety personal

appea

2015 APPEA Health, Safety and Environment Conference

HSR Forum & OSR Forum Perth, Western Australia **7–9 September**

THE UNIVERSITY OF WESTERN AUSTRALIA
Achieving International Excellence

Reducing uncertainty in life extension decisions

David Finch and Melinda Hodkiewicz

THE UNIVERSITY OF WESTERN AUSTRALIA
Achieving International Excellence

David Finch

MMaintMgt, MIEAust, CEng, CMarEng, MIMarEST, CMRP, CAMA

- Independent Consulting Engineer
- Chartered Engineer, Chartered Marine Engineer
- Masters in Maintenance Management
- Post Graduate Diplomas in Asset Integrity Management and Engineering Project Management
- Over 40 years in the Oil & Gas Industry
- Certified Asset Management Assessor

THE UNIVERSITY OF WESTERN AUSTRALIA
Achieving International Excellence

Motivation

Ref: http://www.offshoreenergytoday.com/oil-gas-uk-issues-guide-for-management-of-ageing-offshore-infrastructure/

THE UNIVERSITY OF WESTERN AUSTRALIA
Achieving International Excellence

Status

- Much work done by the HSE - KP4 program http://www.hse.gov.uk/offshore/ageing.htm
- HSE undertook 33 onshore and offshore inspections 2011-2013
  - Safety management systems;
  - Structural integrity;
  - Process integrity;
  - Fire and explosion;
  - Mechanical integrity;
  - Electrical, control and instrumentation;
  - Marine integrity;
  - Pipelines;
  - Corrosion;
  - Human factors.

Selected issues from KP4

THE UNIVERSITY OF WESTERN AUSTRALIA
Achieving International Excellence

- No mature ALE (Asset life extension) policies and procedures
- Observed physical condition of offshore assets combined with very large backlogs suggests problems implementing process
- Obsolescence is an area of concern
- Evidence of missing data, and insufficient data
- Insufficient use of data trending to enable forecasting of when equipment is likely to fail its criteria of non-conformance.
- Insufficient use of data and information, which could lead to a better understanding of the consequences of creeping change.

Key recommendations for industry

THE UNIVERSITY OF WESTERN AUSTRALIA
Achieving International Excellence

- Continue promoting industry-leading ALE management practices.
- Develop and encourage a corporate culture which embeds an ALE philosophy into AIM for the long-term future.
- Improve focus on obsolescence management.
- Early preparation for life extension, enabling the identification of key issues to be managed to assure continued safe operation.
- Incorporate decommissioning into ALE management.
- Develop leading ALE KPIs to help prevent future loss of integrity.
- Undertake periodic ALE management audits.
- Improve data management, eg maintaining complete SCE and SRE lists, and contemporaneous engineering drawings.
- Improve data trend analysis.
- Further promote workforce involvement through Step Change in Safety.
- Further develop approaches to recruit and retain the skilled workforce necessary to sustain the viable long-term future of the offshore oil and gas industry.

THE UNIVERSITY OF WESTERN AUSTRALIA
Achieving International Excellence

Illustrative Asset Life Cycle

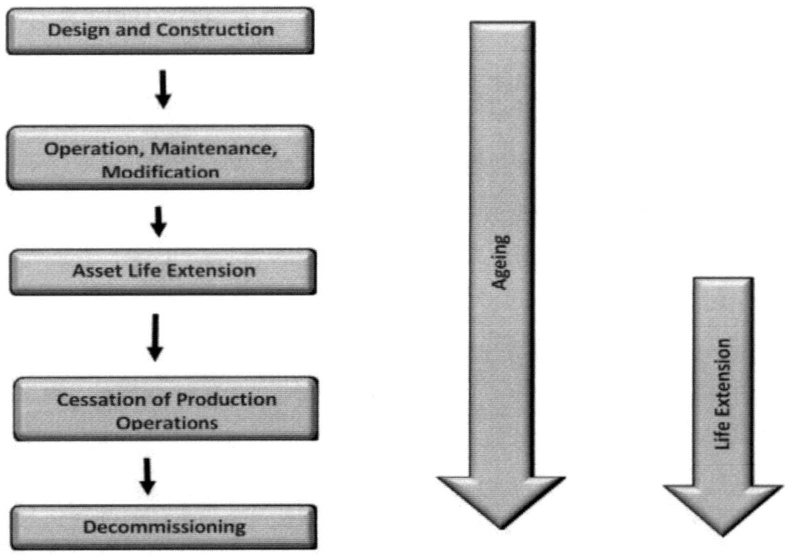

THE UNIVERSITY OF WESTERN AUSTRALIA
Achieving International Excellence

The ageing process can significantly alter the performance, availability and risk profile of a facility. The integrity management of ageing installations depends crucially on knowing equipment condition and associated degradation processes, as well as identifying obsolescent systems or components.

As a result, age-related issues are playing a much greater part in thorough reviews and periodic reviews of safety.

The ageing phenomenon is not necessarily related to the chronological age of equipment, but is a function of its condition and how this is changing.

Equipment which faces harsh environmental, chemical or operational conditions, is often subjected to poor maintenance, or has a high cycle rate and is more likely to deteriorate over time.

Thorough monitoring of plant condition and tuning of maintenance and upkeep requirements is a logical step towards sustaining or improving overall reliability and maintainability.

THE UNIVERSITY OF WESTERN AUSTRALIA
Achieving International Excellence

Degradation

- **Material Properties** - *Is the material choice still appropriate given the effects of ageing?*
- **Operational Conditions** - *Is continuous & cyclic operation for the remaining life of the plant justified? Are restrictions needed?*
- **Maintenance Practices** - *Should maintenance be reactive, preventive, condition-based or a combination?*

Obsolescence

- **Obsolete Equipment** - *Can I get replacement for my 1960s instruments?*
- **New Needs** - *This technology can't handle my new requirements!*
- **New Technology** - *Should I swap this bulky difficult to maintain kit with new technology which is virtually maintenance free?*
- **New Standards** - *I need to upgrade my waste treatment process to meet new requirements.*

THE UNIVERSITY OF WESTERN AUSTRALIA
Achieving International Excellence

Current Maintenance Regime

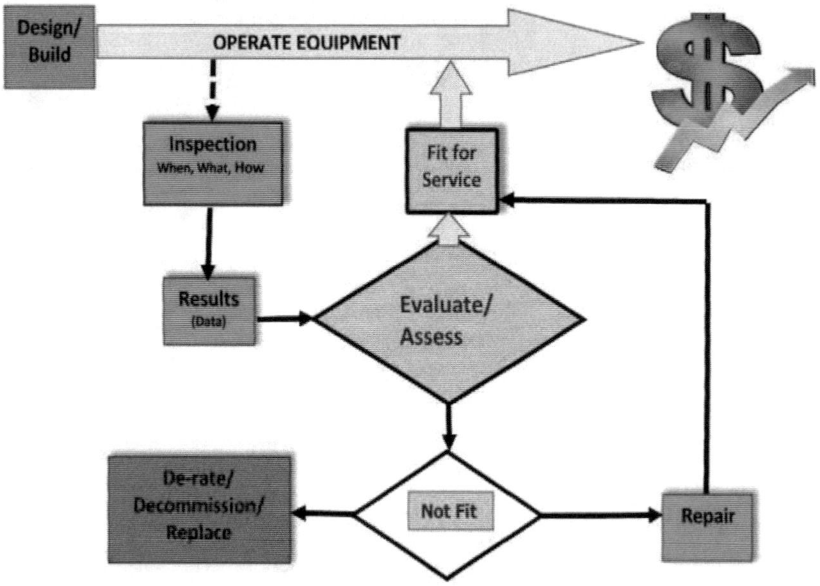

THE UNIVERSITY OF WESTERN AUSTRALIA
Achieving International Excellence

Current Issues to be Addressed?

- Deferred Maintenance/ backlog/ obsolescence
- Periodic Reviews
- Condition Assessment
  - Risk based maintenance
  - Fitness for service
  - Expected life
  - Asset life plans
- Additional operational or performance benefits that can be realized from equipment upgrade or replacement.

THE UNIVERSITY OF WESTERN AUSTRALIA
Achieving International Excellence

Our focus

- Hidden equipment
  - Infrastructure e.g.
    - Lagging
    - Cables & Cable trays
    - Tertiary steel (e.g. handrails, gratings etc.)

- Why?
  - Often overlooked in the budget for ALE
  - Data is usually poor
  - Remediation is expensive
  - Regarded as non-core (production) infrastructure
- KP4 report found that
  - AIM focus has been on the pressure containment envelope, with fabric maintenance often having had little recognition, resulting in a legacy of degradation.

The challenge

- The main challenges for ALE decision making, especially for hidden assets, is …

DATA

- What decision are we making?
- What data is required for the decision?
- How much data do we need?
- Where is it?
- What about the data we don't have (known unknowns)?
- How are we going to use it?

In an ideal world ....

- At ALE time we
  - Would have annual Asset Management Plans through the life cycle of the asset
  - Could then audit the asset(s) and compare findings with what we know from the AM Plan documentation
  - Would have data for models and confidence estimation
  - Would have fewer surprises
- Each year we'd know….
  - What work was done, where and why
  - How much it cost (hence LCC)
  - How the asset had performed
  - How the condition had changed over time
  - How predictions compared to actuals

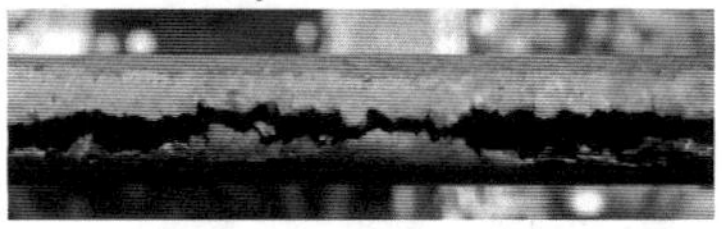

THE UNIVERSITY OF WESTERN AUSTRALIA
Achieving International Excellence

What is an Asset Management Plan?

- An AM Plan is a
  - short document that captures key asset information to inform the annual budget and planning process
  - Is asset specific
  - Pulls data in from existing corporate sources for recording, comparison and trending
  - Only data relevant to decision making is included
  - Should tell you if your operational strategy and maintenance plan is working or not, what risks the asset poses and what its costing you
- Being efficient about AM Plans is a challenge

Example AM Plan – Cable tray

THE UNIVERSITY OF WESTERN AUSTRALIA
Achieving International Excellence

Asset information	
Installed Length of trays	10000m
Addition (current year)	70 m
Average age	30 years
Age of oldest 10%	40 years

Level of service	Measure	Metric	Target	Actual	Status	Data confidence
Production performance	Interruptions	Hrs of downtime	2	0	X	High
Cost	Cost	Cost/ m installed	2000	1500	√	Low
Maintenance performance	PM compliance	% PMs completed/ % scheduled	75	75	√	Medium
	Scheduled replacement	% m of scheduled replacement completed	100%	40%	X	Medium
Condition	Condition index	Condition index	<10% 3 <30% 2 60% = 1	35% 3 45% 2 20% 1	X	Low

Example graphs

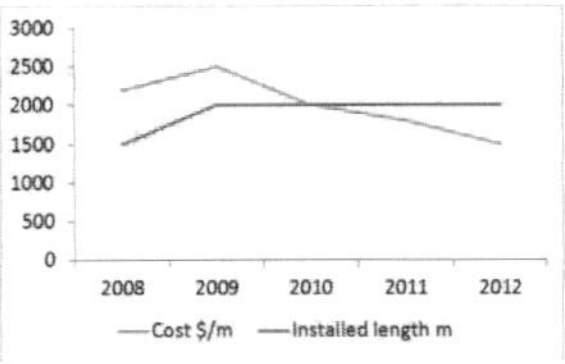

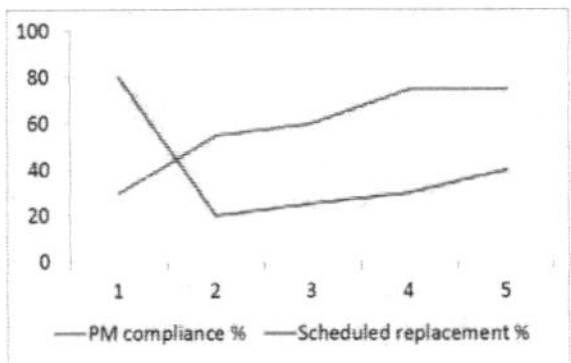

Risk	Ranking

Recommended Actions
Continue with current maintenance strategy of replace 10% of tray in condition 3 each year
Action above maintenance strategy – this requires time from production to do the work
etc

ISO 55001 Asset Management 6.2.2.

THE UNIVERSITY OF WESTERN AUSTRALIA
Achieving International Excellence

When planning how to achieve its asset management objectives, the organization shall determine and document:

- the method and criteria for decision making and prioritizing of the activities and resources to achieve its asset management plan(s) and asset management objectives
- the processes and methods to be employed in managing its assets over their life cycles
- actions to address risks and opportunities associated with managing the assets and how the risks can change with time
- what will be done
- what resources will be required
- who will be responsible
- when it will be completed
- how the results will be evaluated
- the appropriate time horizon(s) for the asset management plan(s)
- the financial and non-financial implications of the asset management plan(s)
- the review period for the asset management plan(s)
- actions to address risks and opportunities associated with managing the assets, taking into account how these risks and opportunities can change with time, by establishing processes for:
  - identification of risks and opportunities
  - assessment of risks and opportunities
  - determining the significance of assets in achieving asset management objectives
  - Implementation of the appropriate treatment, and monitoring, of risks and opportunities.

THE UNIVERSITY OF WESTERN AUSTRALIA
Achieving International Excellence

Recommendations

- ALE should be Data Management, should be done through the life of the asset?
- AM Plans can and should be part of annual business cycle
- AM Plans are a requirement of the new ISO AM Standard
- AM Plans should concentrate on the data you need for decision making
- Aim is to reduce uncertainty in ALE decisions
- It's not too late to start now

Excellence in the Oil and Gas Industry

- Our experience shows that the oil and gas industry is essentially reactive in nature
  - Inherently inefficient
  - Contributes to the continued occurrence of major incidents
    - Generates a culture of work arounds
- We have been involved in assisting more than 250 oil and gas assets around the world
- The next diagram illustrates the data we've compiled

Measured Levels of Excellence

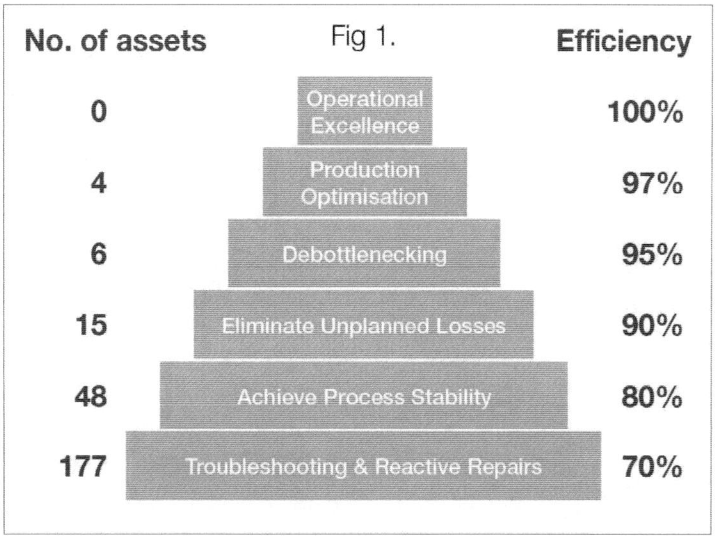

* The results from a sample of 250 assets

Operations Performance

Operational Excellence is an upward spiral of continuous improvement or just as importantly organisational decline is a downward spiral of inefficiency

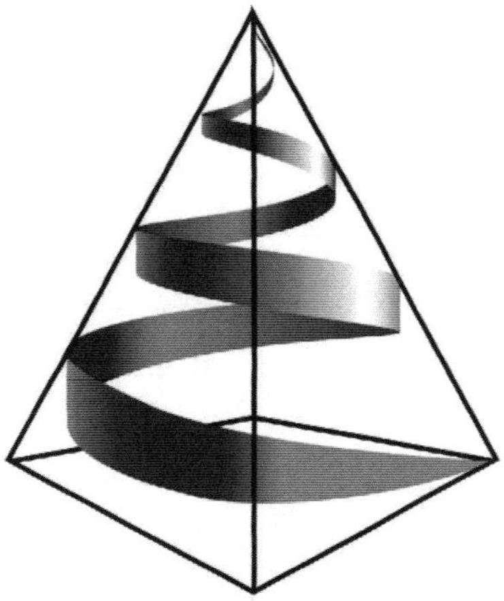

Cross section of the pyramid

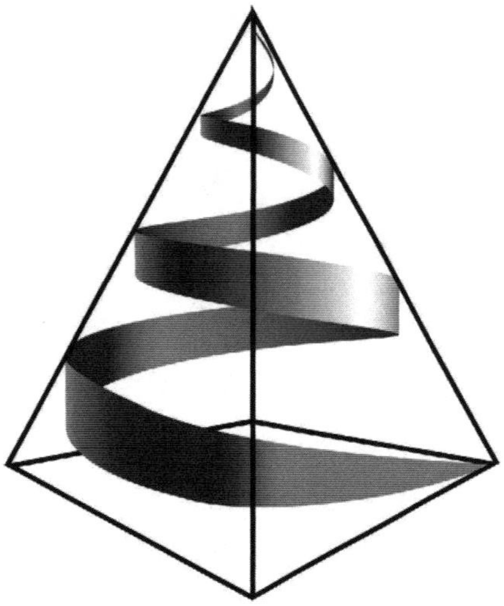

- Represents quantity of work
- Increasing work indicates inefficiency
- Overload drives reactive behaviour

Reactive Behaviour

- Reactive behaviour, or crisis management, inevitably leads to a 'Work Around' mentality
- Examples - Offshore Installation Manager's (OIMs) standing orders or Temporary Operating Procedures (TOPs)
- A prime example is the TOP to handle an alarm flooding problem:
  - 'There will be two panel operators at all times, one to operate the plant and the other to acknowledge and analyse the alarms'
- Or another classic:
  - 'Earthing problems in the accommodation mean there is a risk of electric shock. A fluke is to be attached to the circuit and monitored'

Process Safety in a Reactive Environment

- Proactive improvement is sustained by establishing a culture of root cause analysis and defect elimination
- Industry improvement on slips, trips and falls:
  - Created a culture that focuses on these statistics
  - Developed relevant tools and provided resources
  - Feed the information to the sharp end

- We continue to learn from major process safety incidents
  - Baker report recommended systems of leading indicators
  - Buncefield report promoted IEC61508
  - Flixborough gave impetus to HAZOP
- We have the statistics, tools and resources for process safety but we have not closed the improvement loop
  - We emphasise slips, trips and falls at the sharp end
  - We don't feed practical information on process safety to the sharp end

Sharing Information

- We attend a safety brief on slips, trips, falls and personal safety when we visit an asset for the first time
- Commonly used model to present information is the safety pyramid developed by H.W. Heinrich in 1931

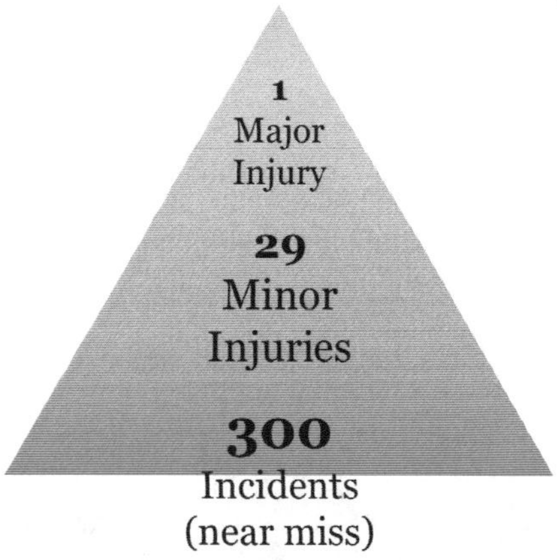

Safety Pyramid

- This model was the basis of Frank Bird's work in 1969 on ratios

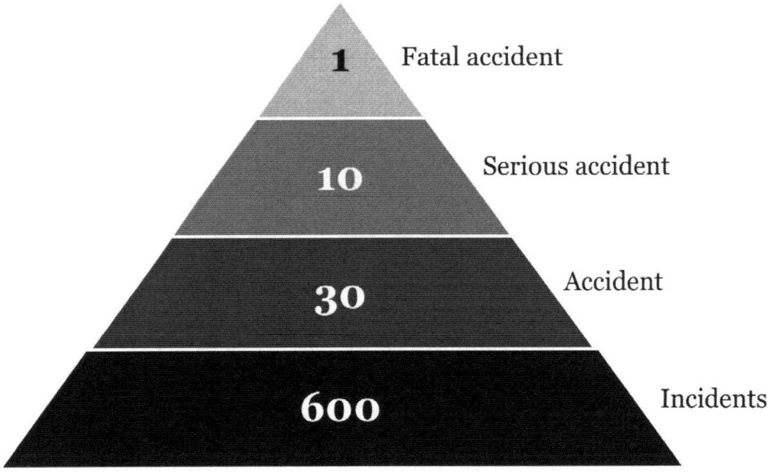

Recent Data Contradicts

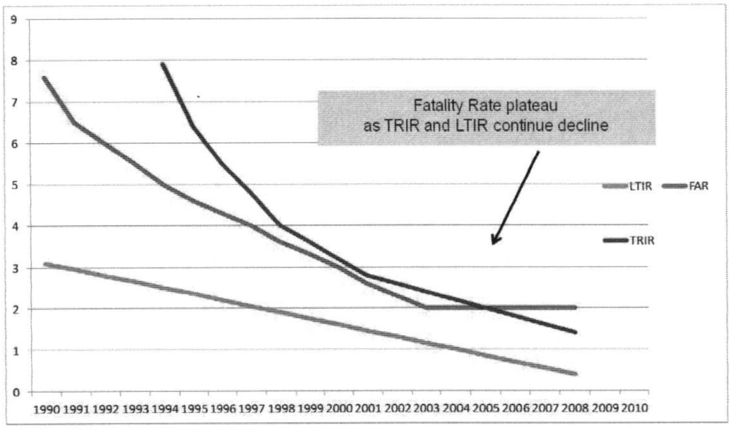

LTIR
Lost Time Injury Rate
per million hrs

FAR
Fatal Accident Rate
per 100 million hrs

TRIR
Total Recordable
Injury Rate
per million hrs

GLOBAL FATALITY RATES ARE NOT IMPROVING

Statistics indicate that the Fatal Accident Rate (FAR) is not directly correlated to Total Recordable Injury Rate (TRIR) and Lost Time Injury Rate (LTIR)

Causal Relationships

- Slips, trips and falls produced a major improvement as the data shows
- Fatality rate is now dominated by major incidents not individual fatalities
- The ratio of potential hazards to first aids has relevance
- The ratio of first aids to major incidents has no relevance
- In our enthusiasm for pyramids and ratios we sometimes confuse causal relationships
- We overcome this problem by utilising a second safety pyramid

≠

Process safety template

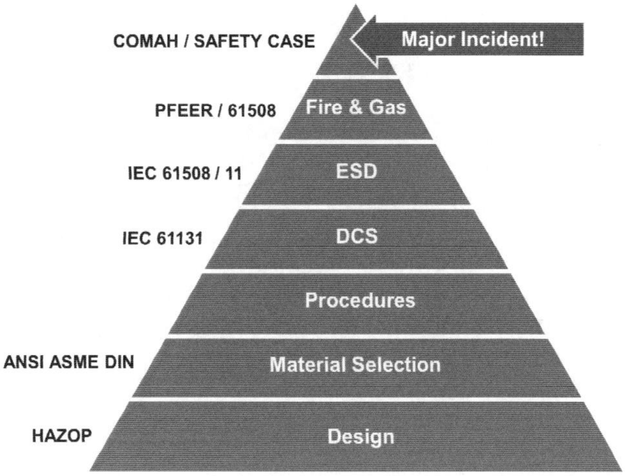

Safety Pyramid

Process Safety Layers

- A metric is selected for each layer of the pyramid
- A design is initially reviewed for safety through an inherent safety review
  - Fallen into disuse in recent years
  - Current trend towards safety by addition demonstrates this review is needed but no easy relevant metric exists
- Design subsequently reviewed for process safety through HAZOP
  - Depending on the age of the plant:
    - Simple check of action closure may reveal process safety has been compromised by failure to adequately complete actions right through to physical modification and successful commissioning
    - An aging plant will benefit from a HAZOP to review the bigger picture created as a result of ongoing process change and modification
  - Relevant metric is HAZOP action closure

Process Safety Pyramid Metrics

- Materials selection: % inspections completed to schedule
- Operations procedures: % procedures signed off after use
- Maintenance procedures: % safety critical procedures completed to schedule
- DCS: Reciprocal % of alarms versus EEMUA 191 targets
- DCS: % control loops in automatic
  - Why are loops in manual accepted as normal?
- ESD: % loops 'live' multiplied by % PPMs versus plan
- F&G: % loops 'live' multiplied by % PPMs versus plan
- All possible metrics for size of holes in Swiss Cheese layers

Measured Information

- Every time we visit an asset we attend a briefing on safety
  - We receive information on 'slips, trips and falls'
- We enter the control room and observe:
  - Control loops in manual
    - Our observed average is 35% of loops in manual
  - Overrides on protective loops
    - Our observed average is 18 overrides per asset
  - Failure to use procedures

- Which is the greater risk?
- Deming's basic tenant is 'What is not measured cannot be improved'
  - We might also add '...and the measures must be shared with the right people'
- The second pyramid provides a model for sharing this information

Hard Truth

- Desire to produce leads people to look for reasons why it is safe to proceed
  - 'It's no worse than anywhere else'
  - Directly opposed to looking for reasons why they should stop
- Bayer Crop Sciences, West Virginia, CSB
  - Desire to start up quickly lead to working around layers of protection
- Buncefield: Alarms used to guide filling, level indication failed
- Texas City: Column filled with known high high level trip failure
- Macondo: Negative pressure test accepted when integrity was unproven

LISTEN UP

Conclusion

- What would happen if teams treated process safety the same way as slips, trips and falls? We'd look for reasons to stop
  - Proven an overview of related metrics enables and empowers cultural change
- Process safety measurement must be available at the sharp end
  - Must not be purely technical realm
- A root cause of major incidents is the inherent, cultural attitude of revelling in the glory of overcoming obstacles (reactive behaviour)
  - Layers of protection can be perceived as obstacles
- The process safety pyramid provides a tool to illustrate (and quantify) the current risk associated with layers of protection designed to prevent major incidents
  - The size of the hole in the layers of the Swiss Cheese

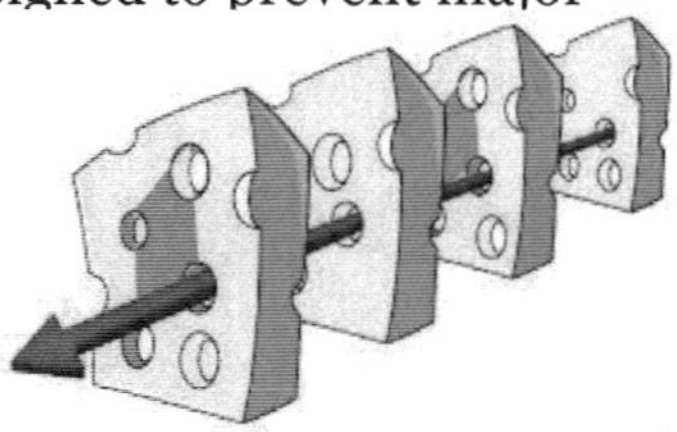

CCPS Vision 20/20
A Call to Action

SHAKEEL KADRI
EXECUTIVE DIRECTOR, CCPS

2015 CCPS ASIA-PACIFIC PROCESS SAFETY CONFERENCE, PERTH, AU
11 FEBRUARY 2015

An AIChE Technology Alliance
CCPS
Center for Chemical Process Safety

Why Create a Vision?

- Our common goal - Reduce incidents
- Why the Vision?
  - We want to reach the goal faster
  - Leverage our collective strength
  - Help CCPS and other organizations identify projects/initiatives to help us reach the goal
  - Make a step change in management of process safety

It's time to leverage our resources, knowledge and skills to all strive for a common goal of great process safety performance. CCPS's Vision 20/20 describes that vision.

Vision 20/20
The CCPS Project

- Envisions the characteristics of companies with great process safety performance in the future, 2020
- Committee members
  - **Cheryl Grounds - Chair**
  - **Dave Jones**
  - **Jack McCavit – CCPS**
  - **Jeff Fox**
  - **Joe Allaben**
  - **Karen Tancredi**
  - **Louisa Nara**
  - **Mike Broadribb**
  - **Pete Lodal**
  - **Samantha Scruggs**
  - **Scott Berger**
  - **Steve Arendt**
  - **Todd Aukerman**
  - **Walt Frank**

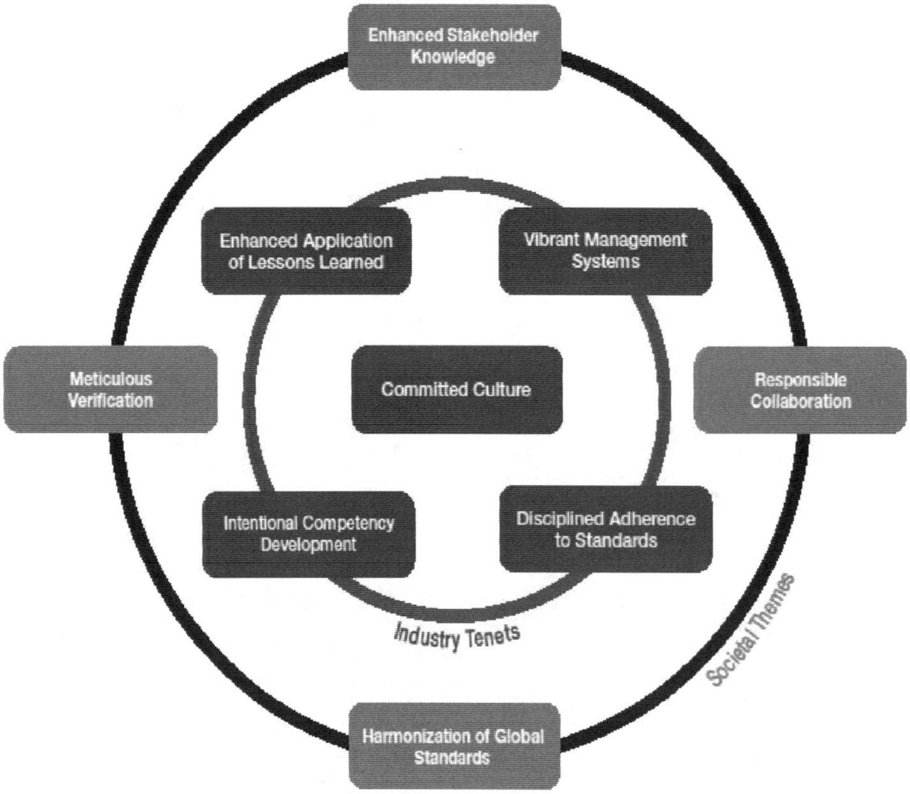

Envisioning the Future…

Company Tenets

1. Committed Culture,
2. Vibrant Management Systems,
3. Disciplined Adherence to Standards,
4. Intentional Competency Development, and
5. Enhanced Application and Sharing of Lessons Learned.

Societal Themes

1. **Enhanced Stakeholder Knowledge,**
2. **Responsible Collaboration,**
3. **Harmonization of Global Standards, and**
4. **Meticulous Verification**

… achieving great process safety performance

Committed Culture

- Felt leadership from senior executives through plant personnel
- Leadership instills disciplined adherence to the policies and procedures
- Strong sense of vulnerability
- What can you do?
  - ☑ Perform all tasks, correctly, every time.
  - ☑ Support your leadership in demonstrating a commitment to process safety.
  - ☑ Recognize that "it could happen here…"

Vibrant Management Systems

- Structure that clearly defines expectations for all systems that impact process safety performance
- Promotes principles of inherently safer design
- Promotes principles of *Guidelines for Risk Based Process Safety* including fit-for-purpose policies and procedures
- What can you do"

☑ Do your part as a member of the team - use the system!

☑ Share your learnings and suggested improvements to the system.

Disciplined Adherence to Standards

- Company and/or industry standards rigorously followed
- System to ensure that existing equipment meets company minimum expectations
- Risk decision making is used when applicable standards do not exist
- What can you do?

☑ Ensure a system exists to effectively use industry and company standards.

☑ Make sure that the system applies to existing equipment.

☑ As a leader, assure standards are followed and respond accordingly.

Intentional Competency Development

- Ensures that employees who impact process safety performance are meeting the technical and cultural requirements of their job
- It is given that companies provide sufficient numbers of employees
- What can you do?
  - ☑ Develop a plan to fill your competency gaps.
  - ☑ When you're outside your area of competency, contact an expert that can help.
  - ☑ As a leader, build and recognize team member competence.

Enhanced Application of Lessons Learned

- Thirst for learning from
  - Incidents
  - Near misses
  - Benchmarking
  - Jobs done well
- Changes are implemented based on the lessons Learned
- What can you do?
  - ☑ Use a significant incident or near miss as an opportunity to encourage and embed learnings.
  - ☑ Take a personal action, or make a change, as a result of the learning. Sharing only makes you feel good, learning makes a difference.

Envisioning the Future…

Company Tenets

1. Committed Culture,
2. Vibrant Management Systems,
3. Disciplined Adherence to Standards,
4. Intentional Competency Development, and
5. Enhanced Application and Sharing of Lessons Learned.

Societal Themes

1. **Enhanced Stakeholder Knowledge,**
2. **Responsible Collaboration,**
3. **Harmonization of Global Standards, and**
4. **Meticulous Verification**

… achieving great process safety performance

Enhanced Stakeholder Knowledge

- Risk literacy
  - Public
  - Government
  - Industry decision makers
- Process safety fundamentals
  - Chemical engineering students
  - Employees
- What can you do?

☑ Support local STEM education, both in high school and university.

☑ Have a conversation about risk with students and other stakeholders within your community.

☑ Continue to build knowledge throughout your career.

Responsible Collaboration

- Technical, government, labor, community, scientific, academic, and industrial organizations work together
- Remove barriers to report incidents
- What can you do?

☑ Recognize the value that others bring to the table and commit to working together toward the common process safety goal.

☑ Be open-minded and focus on that common goal.

☑ Actively participate in or lead joint industry projects.

Harmonization of Global Standards

- Standards can come from many sources
- Standards can be confusing and conflicting
- In the future, work jointly to streamline practices, eliminate redundancy, and cooperatively address emerging issues.
- What can you do?

☑ Recognize the diversity in standards, be open-minded, and work towards harmonization of standards.

☑ Promote harmonization through active participation in standard and practice writing groups.

Meticulous Verification

- Third parties including
  - Public
  - Non-Governmental Organizations
- Evaluate implementation of process safety programs
- What can you do?

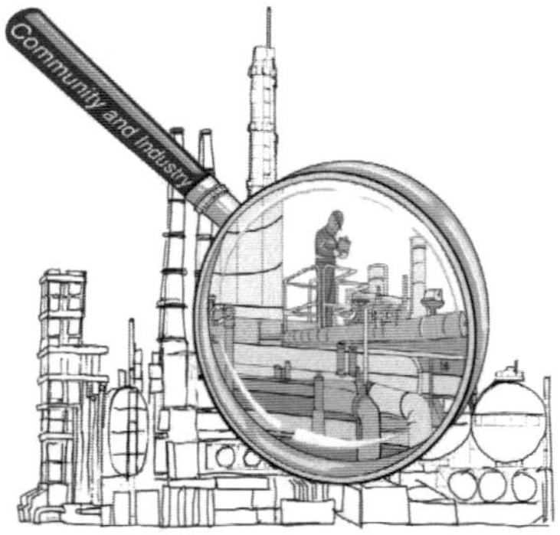

☑ Identify opportunities where third-party verification could add value.

☑ Personally support external verification in your area when it occurs.

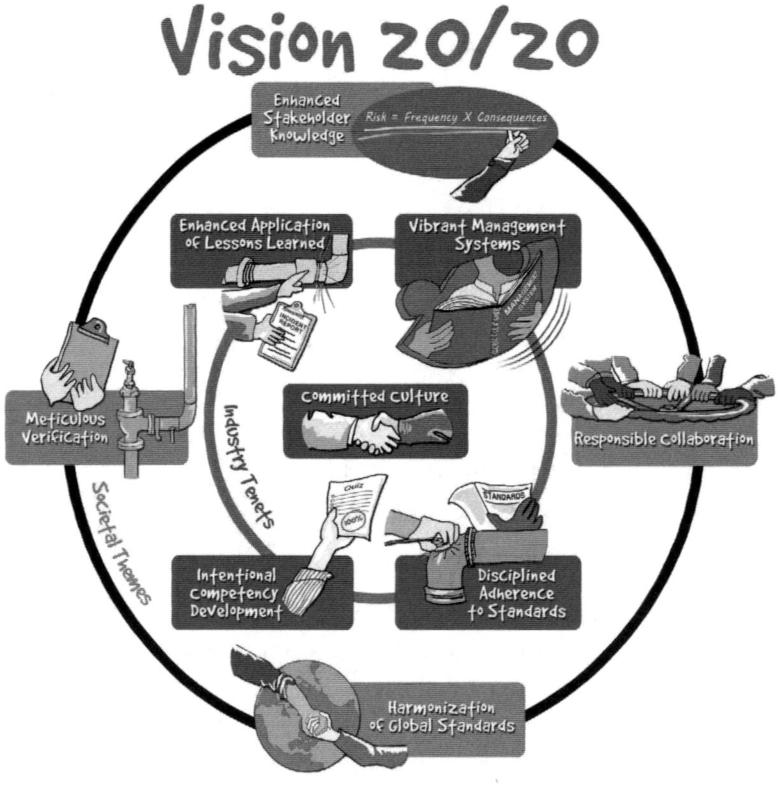

A Call To Action

It's time to leverage our resources, knowledge and skills to all strive for a common goal of great process safety performance. CCPS's Vision 20/20 describes that vision.

- Understand and communicate the five tenets and themes
- Evaluate your contribution
  - within your company, across your discipline industry colleagues, and with your regulatory, academic, and local communities.
- Evaluate performance, seek collaboration and take action

Vision 20/20 Tools - Current

- Brochure
- "A day in the life of"
- One page documents
- Booklets
- Presentations

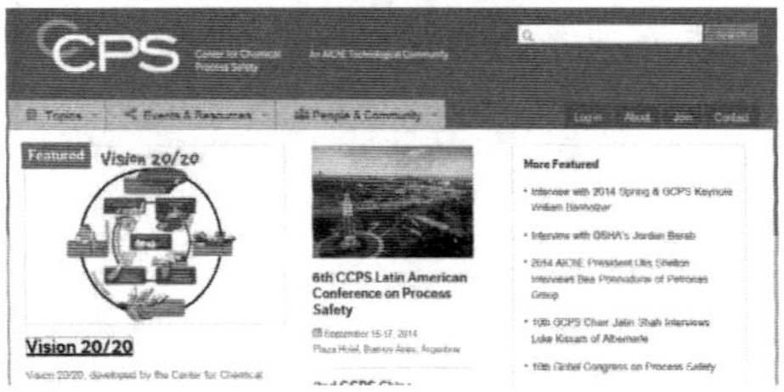

  - Global Congress on Process Safety, New Orleans
  - IChemE Hazards Conference, Edinburg
  - Global Summit on Process Safety – Mumbai
  - Asia-Pacific CCPS Conference – Perth

Vision 20/20 Tools - Draft

- Assessment tool
  - Self assessment by representative team
  - Addresses each industry tenet
  - Intended to be
    - Easy to use
    - Differentiating
- Link to existing resources
  - CCPS texts
  - Worldwide industry resources

On the CCPS homepage

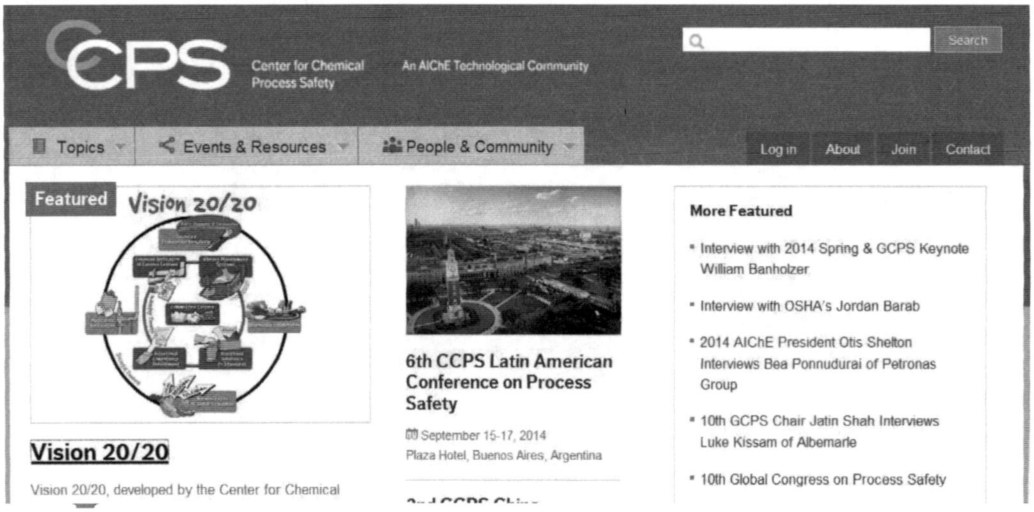

Thank you! Questions?

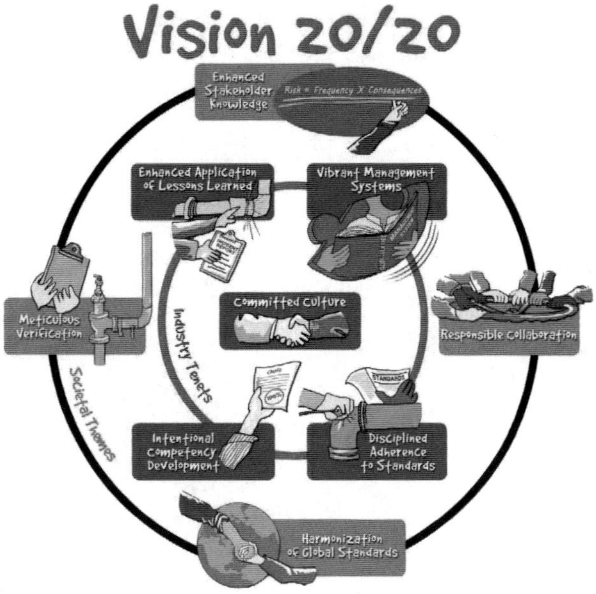

Autonomous platforms for marine environmental monitoring

Iain Parnum, Christine Erbe, Robert McCauley, Alec Duncan, Miles Parsons and Alexander Gavrilov

►► CENTRE FOR MARINE SCIENCE AND TECHNOLOGY   Curtin University

Overview of talk

- Motivation: Environmental Impact Assessment
- Autonomous monitoring of:
  - Underwater noise – stationary logger
  - Oceanography and water quality – real-time and mobile data collection
- Take home messages

The Centre for Marine Science & Technology comprises a multi-skilled group of scientists and engineers committed to the development of technical ocean-related skills in Australia

►► CENTRE FOR MARINE SCIENCE AND TECHNOLOGY   Curtin University

Environmental Impact Assessment

Main requirement for assessment comes from environmental legislation for the protection of marine fauna (e.g. EPBC Act)

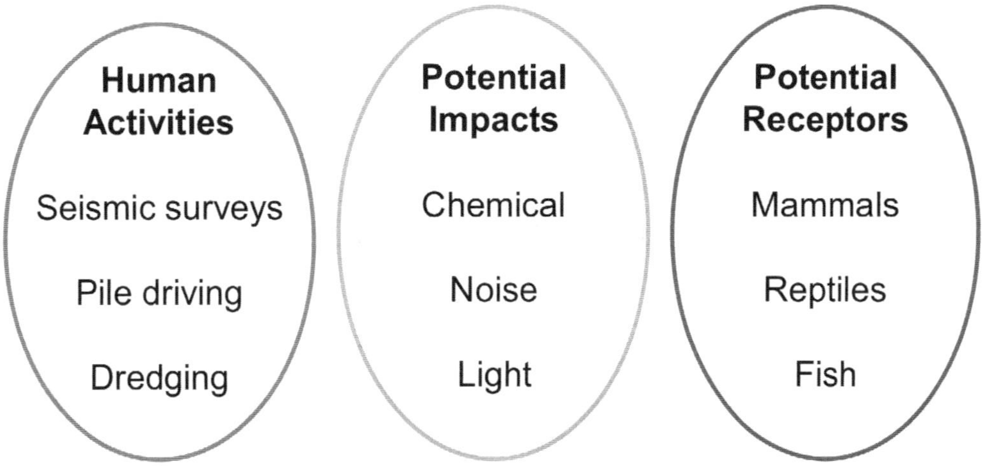

- Establish baseline
- Characterise spatial and temporal variation
- Activity monitored in real-time or post-processed?

Curtin University

Sound in the Ocean

Sound plays an important role in the marine environment as it can travel much further than light

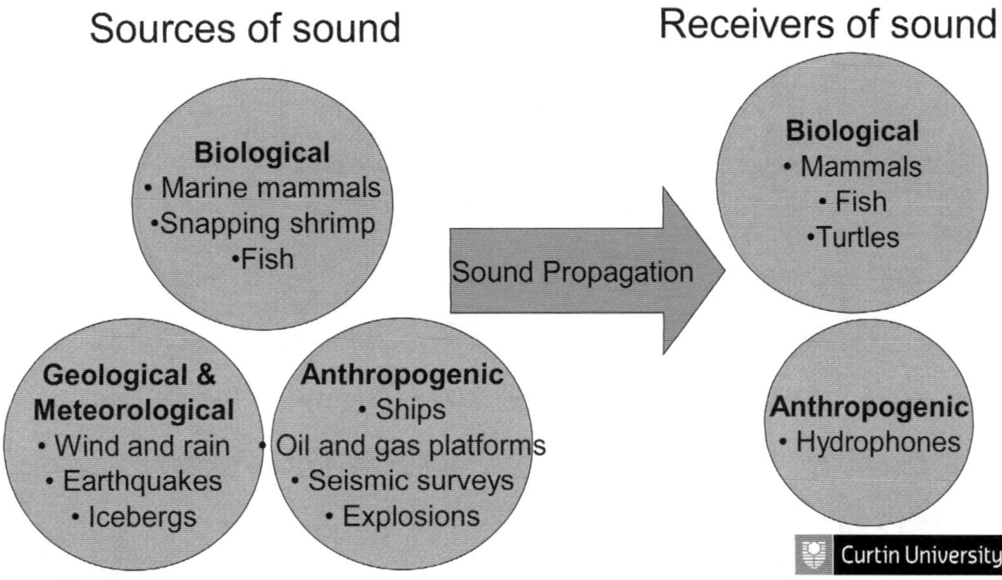

Curtin University

Measuring underwater sound

- Handheld and surface based measurements are suitable for specific events and where spatial coverage is needed
- To characterise temporal variation requires long term deployment, e.g. 6-12 months
- Subsea mooring is best for long term deployments, as:
  - No surface presence
  - Stationary
  - Reduces noise from rain and waves

Example subsea mooring

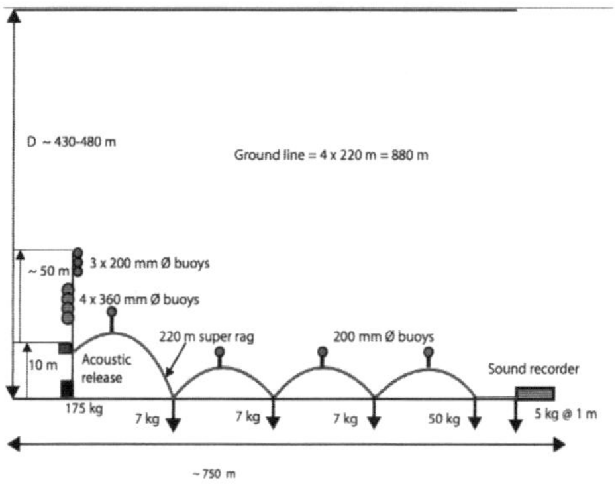

Curtin University

Measuring underwater sound

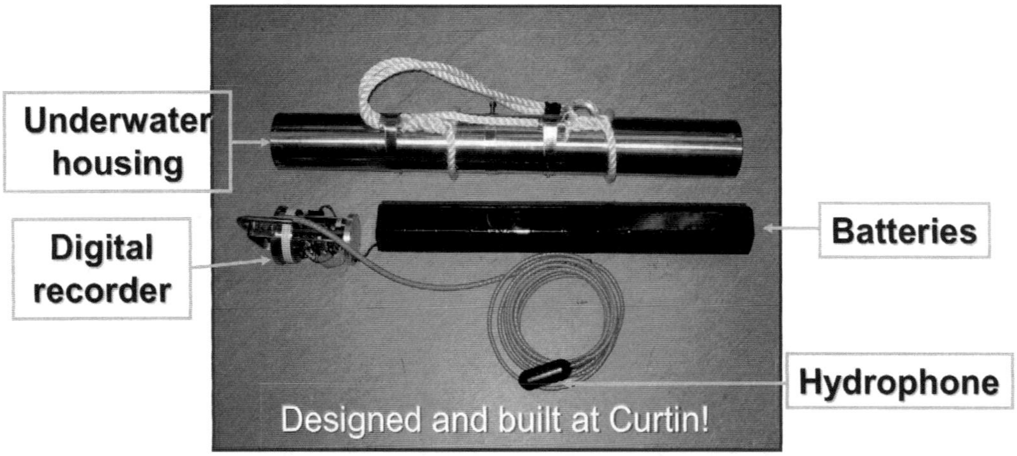

Sea noise data can be recorded over several months (up to 1 year) and then retrieved after recovery of the recorder.

Curtin University

Measuring underwater sound

- Depending on objectives, the recordings can vary in sample rate and recording schedule
- For long term deployments, loggers are typically set to record 5-10 minutes every 15 minutes at a sample rate of 6000 samples/s. For 1 year, this results in:
  - >34,000 files
  - >2,800 hours of recordings
  - >62 billion data points
- Can not review every file -> development of in house software "CHORUS"

Signal waveform

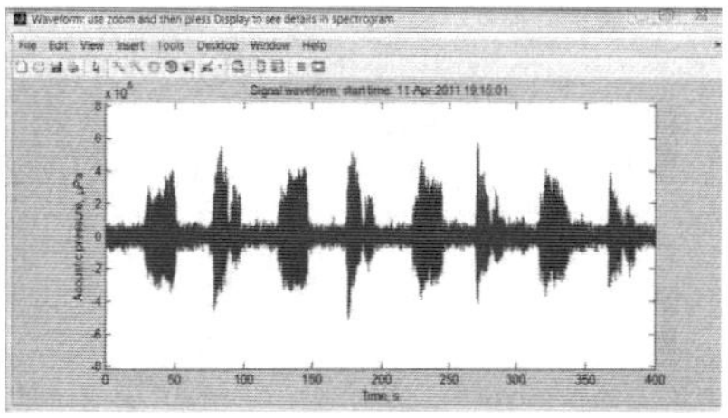

Fast Fourier Transform (FFT)

Spectrogram

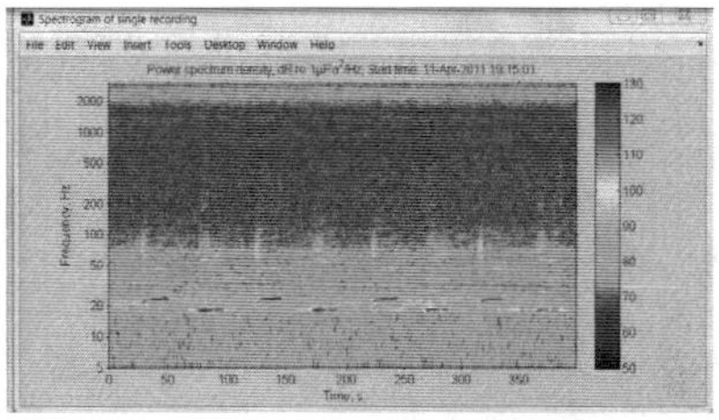

Curtin University

Sound in the ocean

- Averaged spectrogram created to display long term trends
- Algorithms developed to automatically detect specific whale calls

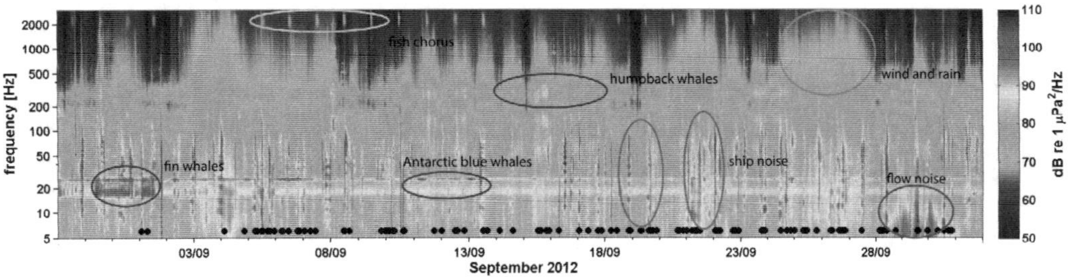

Spectrogram for noise recorded from Perth Canyon during September 2012. The black dots indicate ship passes within 20 km of the sound recorder.

Curtin University

Sound in the ocean

Although data might require disseminating, caution must be applied when trying to simplify the complex.

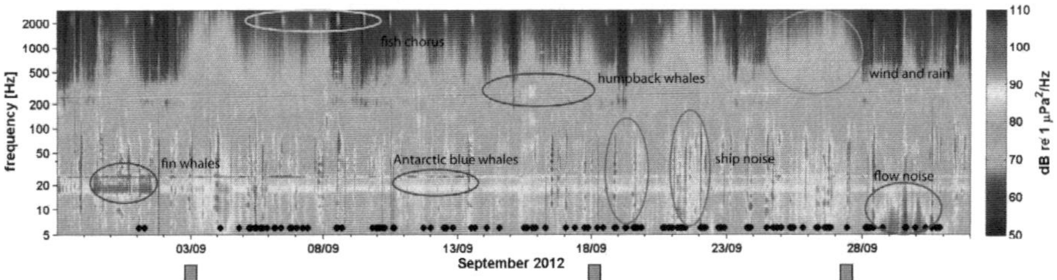

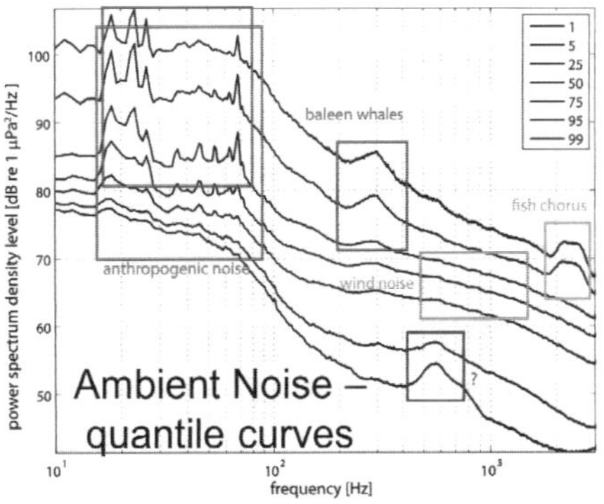

Broadband Sound Pressure Level

- Number and type of whale calls
- Anthropogenic activity

Curtin University

Real-time monitoring: wave gliders

Why real-time?

- Conditions of environmental approval
- For project management

Why a glider?

- Not stationary, so can react to changing conditions or requirements
- Increased safety due to a reduction of personnel in the field

Image of a Liquid Robotics ocean glider showing the surface portion and sub-sea portion with fins, lashed together for transport.

►► **CENTRE FOR MARINE SCIENCE AND TECHNOLOGY**

Curtin University

Wave gliders

- Autonomous:
  - Two part vehicle converts wave motion to thrust
  - Controlled remotely via the secure web
- Vehicle: not a vessel
  - HSE strengths- no fuel on board, no human on board
  - Payloads deployed based on service, on the vehicle or as a tow fish
  - Satellites provide global communication and control
  - Direct measurements of air-sea interface parameters.

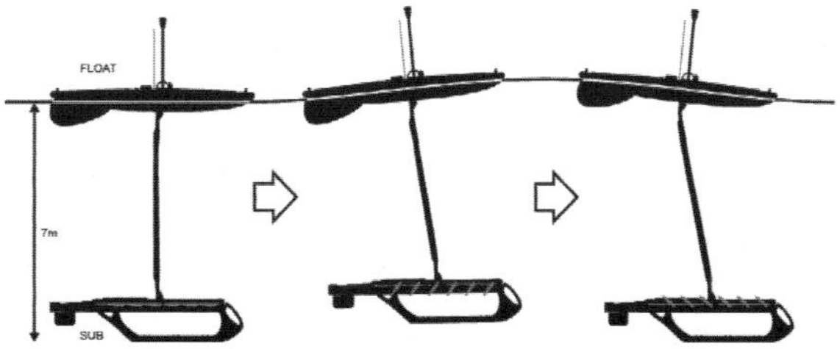

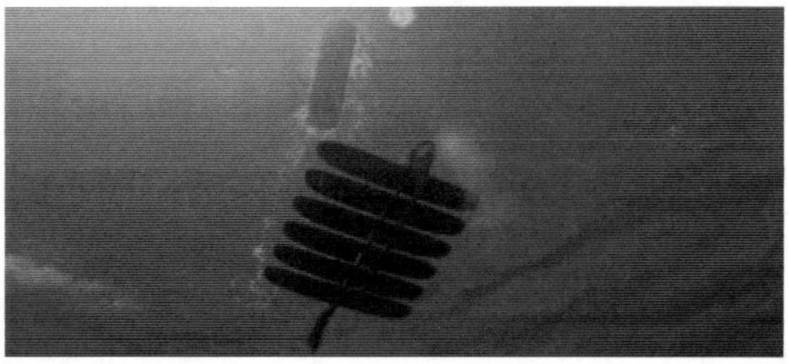

▶▶ CENTRE FOR MARINE SCIENCE AND TECHNOLOGY

Curtin University

Wave gliders: configuration

Sensors attached to the wave glider(s), included:

- **Weather Station** – for air temperature, wind speed and direction.
- **ADCP** – for water current speed and direction, and potentially turbidity.
- **CTD** – for water temperature and salinity.
- **ECOPUCK and CSTAR**– for turbidity.

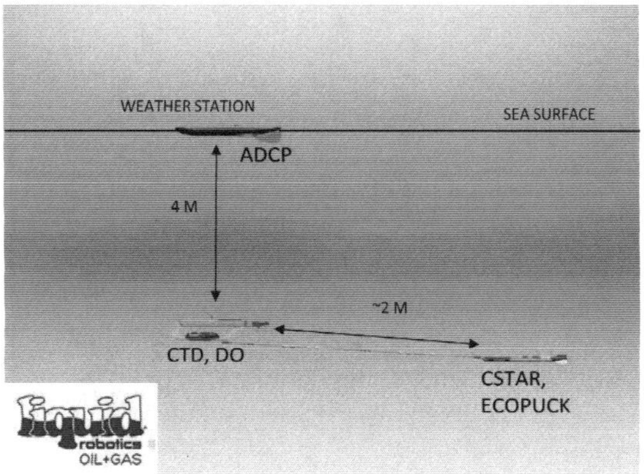

▶▶ CENTRE FOR MARINE SCIENCE AND TECHNOLOGY

Curtin University

Wave glider: turbidity

Why?

- Sediment plumes created from dredging can reduce the light availability
- Real-time monitoring required to check if levels are within what is deemed acceptable

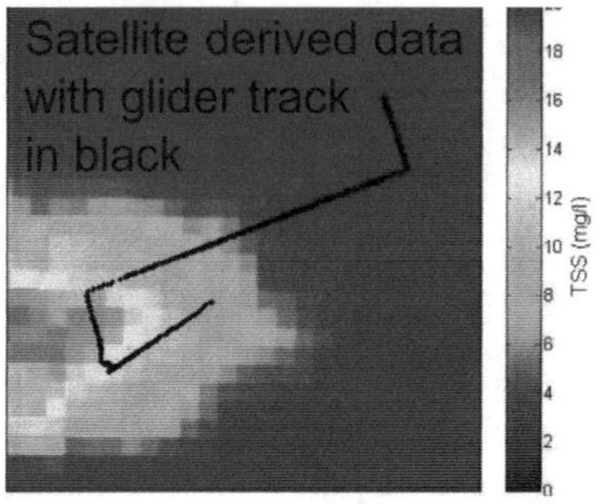

Alternatives:

- Water sampling – increased risk with staff in the field
- Moored turbidity logger – stationary and plumes are dynamic
- Satellite – limited availability and low spatial resolution (~250 m)

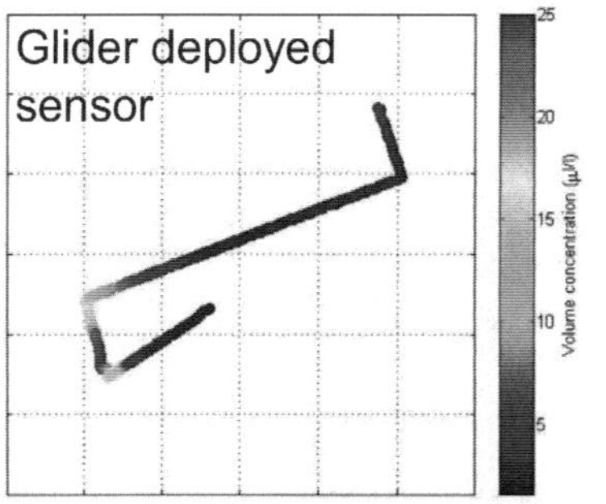

Glider: Water current

The transmitted current data sent by telemetry had a sample interval of 16 minutes (compared to the raw data which is 110 seconds), which was too coarse to see some features, such as the internal wave feature shown below.

Transmitted in near real

Current speed (mm/s)

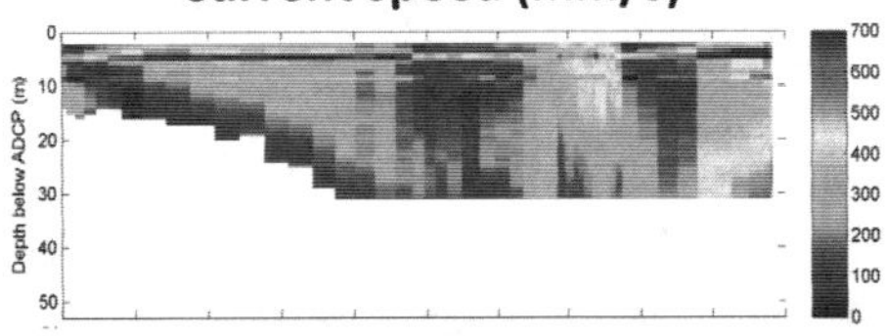

Current direction (deg.)

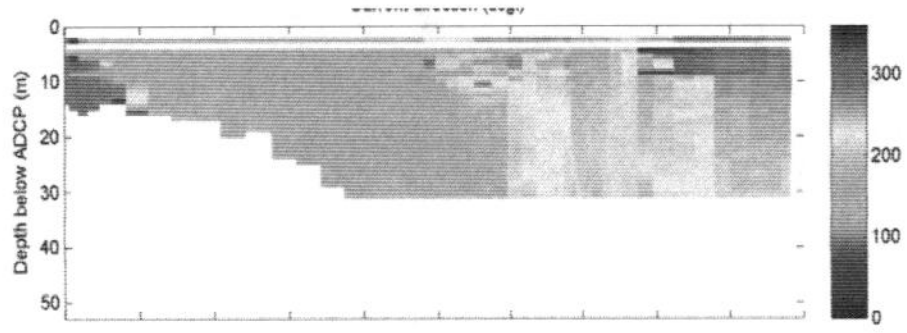

Distance along transect

Logged data

Current speed (mm/s)

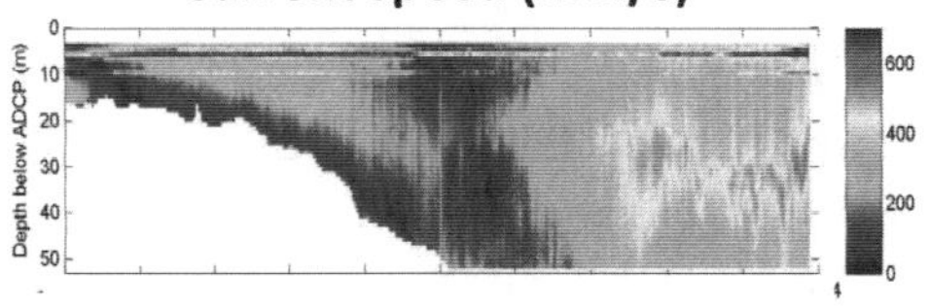

Current direction (deg.)

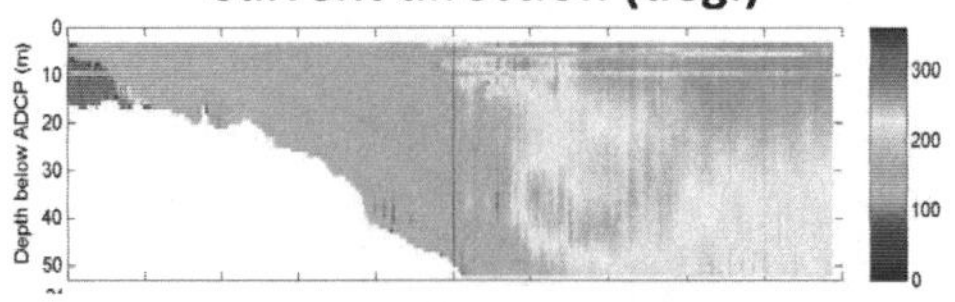

Distance along transect

Glider: Water current

- Near real time transmitted current data captures general trends.
- Below shows the logged values in blue and the 'averaged' transmitted data in red - over all water depths.

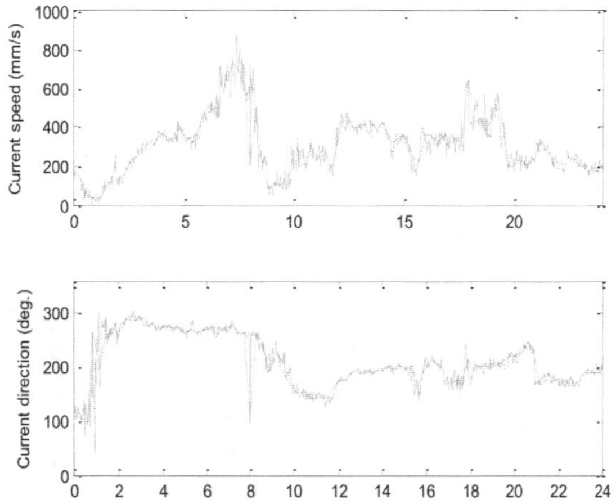

▶▶ CENTRE FOR MARINE SCIENCE AND TECHNOLOGY

Curtin University

Take home messages

Autonomous monitoring requires robust and well tested hardware.

Real time monitoring in the marine environment is possible, but increases costs compared to post processed.

Large data collection requires a well thought out processing pipeline.

▶▶ CENTRE FOR MARINE SCIENCE AND TECHNOLOGY

Curtin University

WorleyParsons

resources & energy

Environmental and social matters

- Why worry about environmental and social considerations?
- Think about your business objectives when investing
  - Maximised access to resources / reserves and production
  - Certainty of project delivery
  - Realising the full value of assets
  - Localisation
  - Future proofed assets
  - Reputation
- Environmental and social performance is key to all of these

Critical to a business (or project)

From objectives to successful outcomes requires effective management of the business / project and its relationships with the 'external' environment

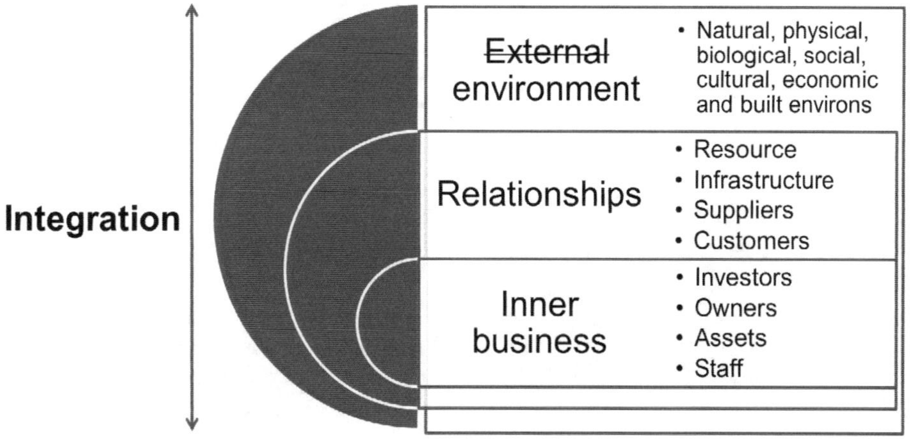

The importance of social licence

- Numerous Australian examples having social licence challenges
  - Coal seam gas development in NSW and Victoria
  - Dredging in Gladstone – fisheries stakeholders / 4 Corners
  - Dredging at Abbot Point – court challenges / 4 Corners
  - Browse LNG Project at James Price Point
  - 100% FIFO for Caval Ridge Mine – Moranbah Action Group
- Conflicts and proxy battles
- Credibility, trust and self-regulation
  1. **Active support**
  2. **Acceptance**
  3. **Tolerance**
  4. **Lawful opposition**
  5. **Civil disobedience**

Different approaches to ESIA

- The different 'personalities' responsible for overseeing ESIA work
  - The minimalist – as little as possible as late as possible
  - The closet greenie – who cares little for business objectives
  - The (political) external affairs manager – beware!
  - The intelligent leaders – still find the work very challenging

Are we regressing?

- Current factors driving us away from an adequate investment in ESIA to achieve business outcomes
  - Shareholder sentiment – uncertain future – limiting appetite for capital investment in greenfield development
  - Plunging commodity prices
  - Dogged focus by management to be austere
  - Political sentiment of 'green tape' reduction
  - Green groups making transparent and collaborative approaches untenable

Objectives of an ESIA

- Objectives of an ESIA
  - Meet requirements of regulators and legal requirements
  - Meet stakeholder expectations
  - Reduce timeframe for securing approvals
  - Minimise need for, and requirements of, supplementary information
  - Inform development of approvals to ensure the project is approved and best possible conditions
  - Inform project planning and create the value that accrues from that (maximised ROI and future proofing)
  - Manage risks and meet company's goals for sustainable development and a positive legacy

ESIA and approvals context

- Can the environment (natural, social, built, economic, cultural) accommodate the development, and if so, subject to what?
- Regulators need to be able to:
  - Review the full impacts of the proposal, including cumulative impacts with existing industry and other significant projects
  - Set conditions which will safeguard environmental and social values
- The project will be measured against principles of sustainability, and application best available technology, where feasible
- Assets need to be secure, in a regional planning context that caters for cumulative impacts
- Need to consider the entire 'asset' (facilities, pipelines, power, water, sewage disposal, roads/rail, community infrastructure, ports,…)

ESIA and approvals - proactive approach

- Get on the front foot
  - Commence ESIA work early
  - Early identification of risks / issues and community views
  - Develop strategies for water management, social impact / opportunities, stakeholder engagement, biodiversity protection, land access, accom.
- Make the investment and effort in stakeholder engagement activities
  - Establish relationships (especially with local communities)
  - If sustained will maintain the social licence to operate these assets
  - Working closely with regulators on constraints and opportunities
- Integrate into engineering design process – use to inform
  - Siting of major infrastructure (using multi-criteria analysis)
  - Basis of design and solutions

ESIA and approvals – proactive approach

- The right team
  - Ensure your consultant is aligned with your expectations
  - Sound leadership required (project and technical) – significant communication exercise
- The right methods
  - Know the local requirements
  - Apply recognised international standards
  - Apply a risk-based approach to asses and manage potential impacts - address priority technical risks / regulator concerns / community concerns

IFC IPIECA ICMM International Council on Mining & Metals ADB Minerals Council of Australia

Use best practice tools to guide engagement

Inform → Consult → Involve → Collaborate → Empower

Increasing level of public interest

IAP2

Creating value

- A proactive approach to impact assessment and approvals will deliver significant value
  - Speed to approvals
  - Understanding and managing risks
    - of the project on the natural and social environment
    - of the natural and social environment on the project
  - Productive stakeholder relationships
  - Long-term approach will provide a level of future proofing
  - Compliance aspects more likely to be manageable

vs

Example - unconventional oil and gas

- The perceived risks of these developments have, in a number of jurisdictions around the world, resulted in:
  - Ongoing and extensive lobby group and protest activity
  - Landowners blocking access to land
  - Landowners and lobby groups taking companies to court
  - Excessive regulatory requirements and moratoriums
- These responses need not be inevitable, but community outrage can be almost impossible to turn around

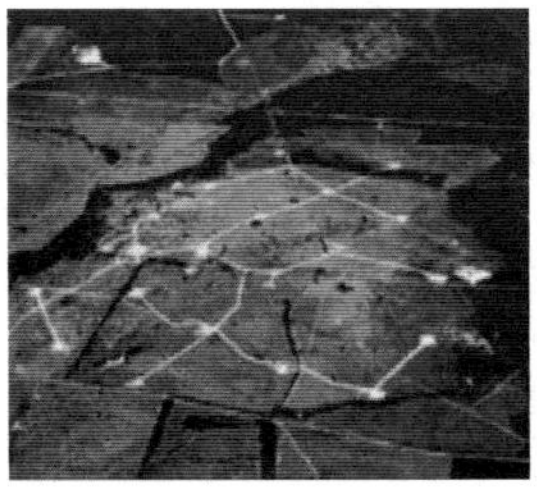

Conclusion

- Environmental and social considerations are critical to meet business / project objectives
  - You need to understand and manage the nature of the relationships of the project with the (natural, social, built, etc.) Environment
  - It takes leadership to be successful, especially during leaner times
- There are significant risks and complexities
  - People tasked to manage an ESIA bring a range of (unhelpful) biases
  - Proxy battles playing out to be cognisant of
  - Runaway community outrage
- Compelling case to be proactive
  - Identify and manage risks
  - Value of considering external risks and opportunities on the project (not just of the project on the environment)
  - Effective stakeholder engagement
- Opportunity for competitive business advantage

WorleyParsons

resources & energy

For more information

Tasman Graham
Director – Environment & Society
tasman.graham@worleyparsons.com
+61 (0)427 236 911

EcoNomics

Process Safety Management [PSM] Metrics

SHAKEEL KADRI
EXECUTIVE DIRECTOR, CCPS

2015 CCPS ASIA-PACIFIC PROCESS SAFETY CONFERENCE, PERTH, AU
11 FEBRUARY 2015

Presentation Outline

- Process Safety Management [PSM] Metrics
- PSM Metrics Implementation at Air Products
- Leading Indicators – CCPS Project
- API/AFPM – Industry Analysis "Deep Dive"

Process Safety Management Metrics

- ✓ Leading Metrics – a forward looking set of metrics which indicate the performance of the key work processes, operating discipline, or layers of protection that prevent incidents.
- o Lagging Metrics – a retrospective set of metrics that are based on incidents that meet the threshold of severity that should be reported as part of the industry-wide process safety metric.
- ✓ Challenges to Safety Systems, Near Miss and other internal Lagging Metrics – Although these events lagging indicator events, they are generally considered to be a good indicator of conditions which could ultimately lead to a more severe incident.

API-754 PSM Metrics

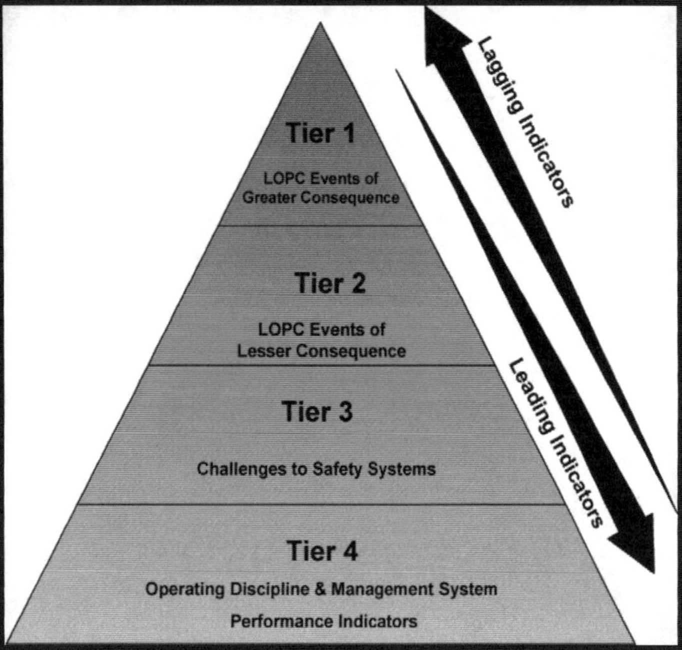

Air Products Related slides have been deleted from this presentation

Process Safety
Leading Indicator Metrics
A CCPS Survey Project

Leading Indicator Metrics Project Team
Team Chair: Kenan Stevick

Author(s)

Sanjay	Choudhary	Tata Chemicals Ltd.	sgchoudhary@tatachemicals.com
Elroy	Christie	Honeywell	Elroy.Christie@Honeywell.com
Jeffrey	Fox	Dow Corning Corp.	jeff.fox@dowcorning.com
Eric	Freiburger	Praxair, Inc.	Eric_Freiburger@Praxair.com
R. Wayne	Garland	Eastman Chemical Co.	wgarland@eastman.com
Andrew	Goddard	Arkema	andrew.goddard@arkema.com
Shakeel H.	Kadri	Air Products & Chemicals, Inc.	KADRISH@airproducts.com
Darrell	Loewe	Formosa Plastics Corp., USA	DarrellL@ftpc.fpcusa.com
Steve	Marwitz	Formosa Plastics Corp., USA	SMarwitz@ftpc.fpcusa.com
Louisa A.	Nara	CCPS	louna@aiche.org
Tim	Overton	TOPS Consulting LLC	topsconsulting@hotmail.com
Jeff	Philiph	Monsanto	jeffrey.e.philiph@monsanto.com
Philip	Rasch	BASF	Philip.rasch@basf.com
Daniel E.	Sliva	CCPS	sliva@capital.net
Kenan	Stevick	Dow Chemical	KPStevick@dow.com

Presentation Outline

- Purpose / Intent
- Result Summary
- Data
- Leading Metrics Visibility
- Additional Guidance needed
- Barriers to Implementation
- Metrics Changed, New Metrics considered
- Next Steps

9TH GLOBAL CONGRESS ON PROCESS SAFETY

Purpose / Intent

- **Purpose:**
  - Provide an update on the use, direction, and effectiveness of Process Safety leading indicators in the chemical and petroleum industries.

- **Strategic Alignment / Intent:**
  - Leading Indicators are intended to help drive performance improvement and lead to reduction in the number and severity of process safety incidents. They do this by identifying improvement opportunities in the management systems for Layers of Preventative defense against Tier 1 and Tier 2 events.

Result Summary

- High percentage of companies recognize the value
- Industry is still 'experimenting and discovering' as to which indicators make the most sense
- Three different focus areas or approaches were identified as most effective in improving performance. These are:
  1. PSM Compliance
  2. Learning Experiences and Management of Deviations
  3. Management Engagement

CCPS An AIChE Technology Alliance Center for Chemical Process Safety

Result Summary

1. PSM Compliance - Follow-up on actions across the spectrum of Process Safety Management Systems:
   - Audit Corrective Actions
   - PHA Actions
   - Completion of Safety Critical Equipment Inspections or Calibrations
   - MOC Actions
   - Unplanned Event Corrective or Preventive Actions

9TH GLOBAL CONGRESS ON PROCESS SAFETY

CCPS An AIChE Technology Alliance Center for Chemical Process Safety

Result Summary

2. Learning Experiences and Management of Deviations
   - Process Safety Near Miss Reporting including fires
   - Challenges to Safety Systems in general and specifically calling out: Safety Instrumented Systems and Relief Device Activations

9TH GLOBAL CONGRESS ON PROCESS SAFETY

Result Summary [3]

3. Management Engagement
   - Picking the most pertinent measures to your operation and getting them in front of leadership;
   - Including Leading Indicators in agendas of various operational reviews and ensuring action.

Data Summary

- 43 companies responded to the survey; 41 Companies [95%] responded that they use leading indicators.
- The use of leading indicators varies by company, from a low of 3 leading indicators to as many as 28 leading indicators for a company.
- Each of the 25 leading indicators was used by one or more of the responding 41 companies.
- 20 companies [45% of responded to the survey] are using 12 or more leading indicators
  - 5 Tier 3 leading Indicators
  - 7 Tier 4 leading indicators

of Companies using a Specific Indicator

All 25 leading indicators were used by one or more of the responding 43 companies.

12 or more leading indicators were used by 20 or more of the 43 companies, (45%)

The red box on the chart highlights the 12 leading indicators used by the 20 or more companies

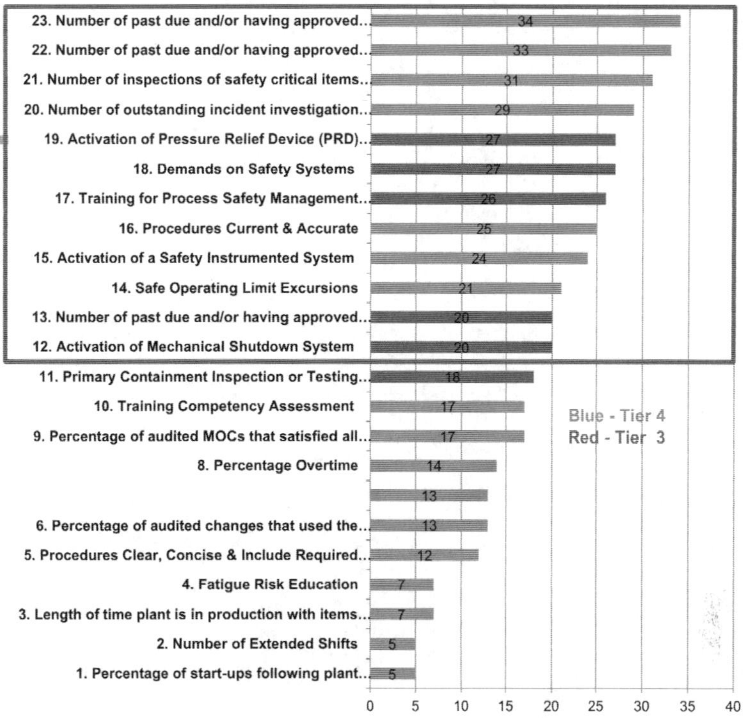

20 Companies or more Using these 12 Leading Indicators

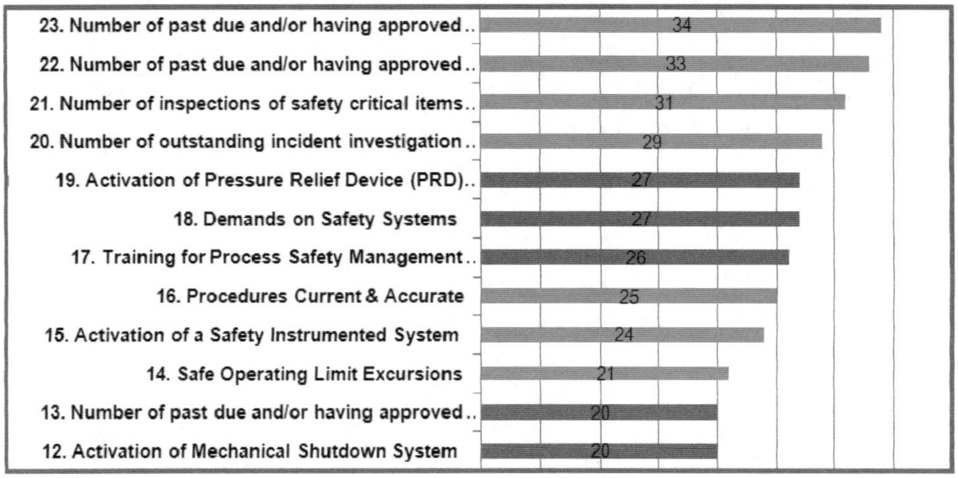

Frequently used Leading Indicators

Tier 3

1. Activation of Pressure Relief Device (PRD) Not Counted as a Process Safety Incident (PSI) or Loss of Primary Containment (LOPC)
2. Demands on Safety Systems
3. Activation of a Safety Instrumented System
4. Safe Operating Limit Excursions

Tier 4

1. Number of past due and/or having approved extension of audit action items
2. Number of past due and/or having approved extension of PHA action items
3. Number of inspections of safety critical items of plant and equipment due during the measurement period and completed on time
4. Number of outstanding incident investigation action items closed
5. Training for Process Safety Management (PSM) Critical Positions
6. Procedures Current & Accurate
7. Number of past due and/or having approved extension of regulatory issue

CCPS An AIChE Technology Alliance Center for Chemical Process Safety

Leading Metrics Visibility

Most companies indicated:

- Roll their measures up into meaningful scorecards:
  - prompting management action,
  - review with senior leadership, (in some cases all the way up to board members)
- Publish data on internal review reports, websites and in newsletters for communication and action across the company

9TH GLOBAL CONGRESS ON PROCESS SAFETY

More Guidance Needed

- ✓ 15 companies felt that the existing guidance was sufficient.
- ✓ 8 companies indicated they had been using these metrics and were comfortable in their understanding.
- ✓ 3 companies indicated that they were not yet using these metrics.
- ✓ 7 companies felt additional guidance would be useful to ensure that "challenges to safety systems" are used consistently within and between companies.

Barriers to Implementation [1]

- Senior Management Commitment/support is essential for the implementation and sustainability of a successful metrics program.
- There can be differences in understanding metrics definitions across the company, e.g., different geographies, acquisitions
- Resources are needed in order to report metrics in a timely manner.
  - Maintaining trained resources who understood the definitions and how to extract the data from the computer tracking system presented a challenge due to transfers, turnovers and retirements.

Barriers to Implementation [2]

- Data Collection Systems
  - Data collection systems often do not readily produce the information companies want to track
  - Significant time and money are needed to develop information relating to leading indicators
  - Computerized / automated data collection systems are helpful in reporting in a timely manner
  - Communication to the workforce about the criteria for inclusion can take a great deal of time and effort
  - Employees need to hear that reporting is important, data consistency will not be perfect from the start but will improve over time

Metrics Change, Discarded and New Metrics being Considered

- 12 companies said they changed or discarded metrics
  - No longer represent areas where improvement needed
  - Emphasis changed to better align with underlying issue
  - Improved performance moved some to background
- 29 companies stated they were considering new metrics. Examples:
  - Alarm management
  - Senior leadership time in the field
  - Fatigue risk management
  - Process Safety Culture
  - Monitor procedures and Incident investigation quality

Next Steps

The following activities are scheduled to continue the focus on the development of leading indicators in the chemical industry:

- Publication of the White Paper in CCPS website for comments after it gets presented in the 9th GCPS
- Communication to CCPS Member companies to adopt leading indicators and inform CCPS of the list of indicators adopted along with the definitions and examples.
- The members of this Project Team to share their own company experiences in use of leading indicators.
- The review to be included in the 2014 Global Congress on Process Safety.
- Biannual Survey from the CCPS members

API related slides have been deleted from this presentation

QUESTIONS??

ENGINEERS
AUSTRALIA

DEALING WITH ZOMBIES: THE ROLE OF PROFESSIONAL SOCIETIES IN ENSURING PROCESS SAFETY

Francis Norman
2015 Engineers Australia Western Australia President

presentation to AiCHE, Perth 2015

Overview

ENGINEERS
AUSTRALIA

- Introduction
- Engineers Australia overview
- Where are the Zombies...
- The role of Professional Societies
- Where to from here

Engineers Australia – What we are

- A member organisation of over 100,000 representing all levels of engineering personnel
- Incorporated by Royal Charter
- Signatory to Washington, Sydney & Dublin accords
- 9 Colleges, each representing a different technical field
- Divisions in all states and territories

Engineers Australia – What we do

- Deliver Continuous Professional Development
- Accredit all Australian University Engineering degrees to international standards
- Standards development
- Administer stage 2 competency recognition (Chartered accreditation)
- Skills assessment for prospective migrants
- Public advocacy
- Established Code Of Ethics (COE)

Code of Ethics

ENGINEERS AUSTRALIA

- Code of Ethics applies to Chartered members, it is part of the demonstration that they are competent to practice independently, that they know their limits and abilities.
- Once Chartered, members commit to maintaining a level of CPD through various forms of ongoing professional development, including within their core skill base, risk management and leadership skills – this is audited every 5 years

ENGINEERS AUSTRALIA

ENGINEERING CAREER PATH

Making a competent (and safe) Engineer

ENGINEERS AUSTRALIA

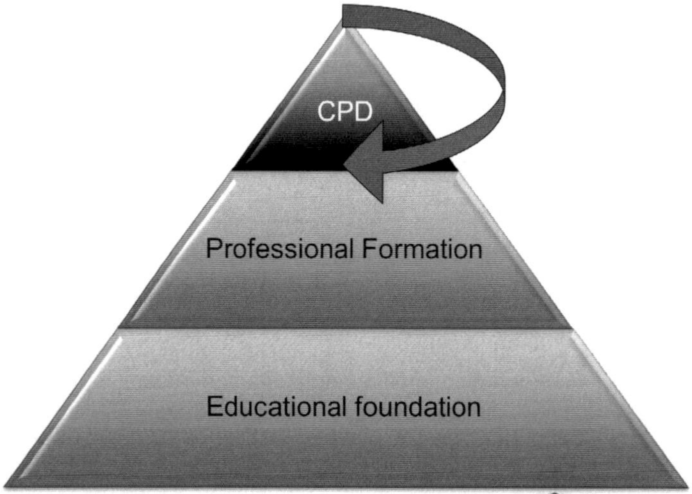

Why do professional societies matter to safety

ENGINEERS AUSTRALIA

- Professional Societies;
  - Set the standards
  - Accredit the Universities
  - Seen by many as the keepers of knowledge
  - Coordinate some research
  - Accredit engineers to practice and maintain currency
  - Positioned to act independently and impartially
- But
  - We could do a lot better…

Why do we have standards

ENGINEERS
AUSTRALIA

Near miss earns engineering firm fine

Friday, 6 February 2015

An engineering company in Western Australia has been fined $16,000 and $5000 in court costs for failing to design a crane to the appropriate Australian Standard, breaching mine safety regulations.

Civil engineer Wally Etschmann of Etschmann Engineering was found to be responsible for the 2013 collapse of an 80 t crane at the Karara iron ore mine in WA. The crane was undergoing load testing at 110% of its capacity when the boom folded in half and dropped the load five storeys to the ground, according to *The West Australian*. Consequently, a building was damaged with the crane narrowly missing workers.

Etschmann pleaded guilty in the Perth Magistrates Court to failing to ensure the equipment complied with the Australian Standard after modifying the crane design by reducing the thickness and weight of boom support plates.

The proceedings marked the first time the state's Department of Mining and Petroleum (DMP) has prosecuted a crane designer, according to DMP state mining engineer Andrew Chaplyn.

"Safety in design is one of the major areas of concern identified by the department," he said. "We hope this sends a strong message to designers working across the mining industry, but in particular in Western Australia, that negligent or inferior services will be not be accepted."

Chaplyn said the DMP was promoting a risk based approach to safety that employs a hierarchy of control.

"Engineering sits above the two lowest levels of control, administrative and personal protective equipment, these two being relied on far too much in WA mining," he said.

Less than two months after the crane collapse, another incident at Karara left a worker with facial injuries. The DMP began investigating crane operations and other issues across mine sites in the state's mid-west in the wake of the two accidents.

Photo: industrymayhem.com

This news item has been posted by Engineers Media, a wholly owned subsidiary of Engineers Australia.

Why do we have standards

ENGINEERS
AUSTRALIA

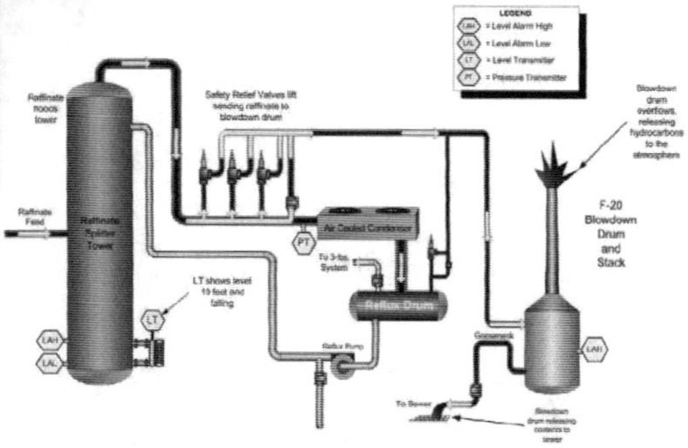

Why do we have standards

- **Safety knowledge & expertise**
  - A system should be developed and implemented to ensure that its executive management, its refining line management above the refinery level, and all refining personnel (including managers, supervisors, workers, and contractors) possess an appropriate level of process safety knowledge and expertise.

Recommendation from the findings into the 2005 Texas City refinery incident

ENGINEERS AUSTRALIA

Where are the Zombies

ENGINEERS AUSTRALIA

Engineering Zombies

ENGINEERS AUSTRALIA

- Hold professional academic qualifications but do not maintain currency
- Claim to be “professionals” and expect to be treated as such, but do not contribute to their profession
- Without currency, these Zombies
  - propagate old and bad practices,
  - make poor mentors for others,
  - hold back progress in evolution of best practice

Competent Versus Zombie…

ENGINEERS AUSTRALIA

Competent

- Demonstrated commitment to CPD and COE
- Judged by peers as able to work independently
- Competencies independently verified to a recognised standard
- Work to their level and area of expertise
- Undertake career path planning

Zombie

- No commitment to CPD and COE
- Judged by self as able to work independently
- No demonstrated or verified competencies
- Work as allowed by employer or society
- No determined career path planning

Reanimation

- The biggest challenges with Engineering Zombies are;
  - **Identification** - picking a Zombie out in a crowd can be difficult but they will eventually show themselves.
  - **Reanimation** - demands that we offer them an alternative to the life of the undead... through a range of attractive professional development opportunities or through regulation

Managing the zombies

- Attracting Zombies to join the relevant professional body and participate, i.e. Charter and maintain CPD
  - How do we get them on board
    - Attraction through better offerings
    - Regulation through government
    - A combination of the above
  - Attraction is typically better than regulation BUT to be attractive we must be seen to have a place.

Pull versus Push

ENGINEERS AUSTRALIA

Pull

- Provide an imperative to be "part of" the profession rather than feed off it
- Offer meaningful CPD
- Offer networking and career advice
- Be seen by employers and society as so relevant the Zombies "must" join

Push

- Establish a set of regulations that mean a Zombie can not have the job prospects they want without joining in with their profession

Pull

ENGINEERS AUSTRALIA

- To pull engineers to give up their zombie lifestyle, professional societies must offer meaningful "value" for membership;
  - Represent the profession so that membership is seen as prestigious and valuable
  - Represent the membership to the broader community on key issues
  - Position membership so that it is seen by employers as essential rather than a nice to have – think medical, legal and accountancy etc.

The role of professional societies

ENGINEERS AUSTRALIA

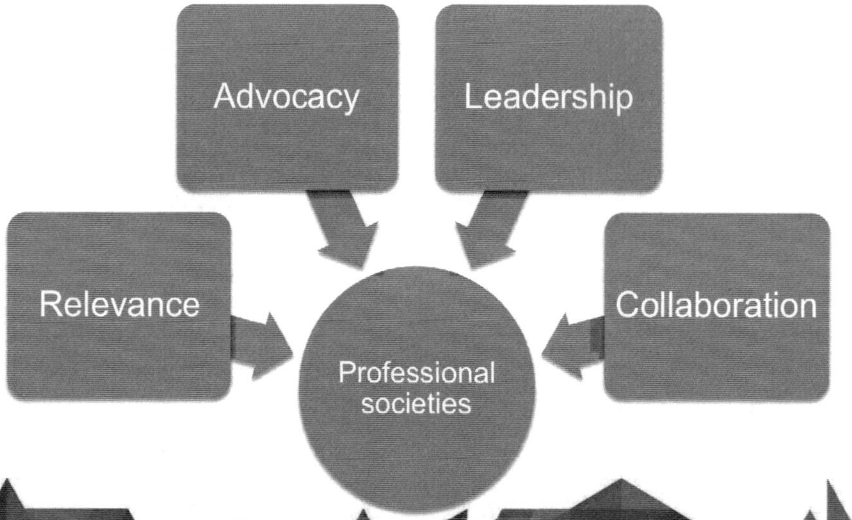

Relevance of Professional Societies

ENGINEERS AUSTRALIA

- **To society as a whole** – we have an obligation to provide impartial advice and represent engineering to the informed and uninformed members of society
- **To our membership** – represent our membership and others in the profession externally and provide relevant career development opportunities
- **To employers** – provide independent verification of skills of personnel, verify qualifications and support the development of standards

Advocacy

- Represent our profession to;
  - **General society** – Participate in discussions that matter to society, don't just be a voice when there is a press event
  - **Government** – Be the trusted voice of our profession to government
  - **Other professional bodies** – Collaborate on positioning on key issues and standards

Leadership

- Provide leadership on key issues;
  - To community
  - To membership
  - To education
  - To government
  - To legislative groups
  - To industry

Collaboration

ENGINEERS AUSTRALIA

- Collaboration between professional societies could be better…
  - We tend to get absorbed in our own internal issues and challenges
  - Try to resolve issues internally
  - Not talk as openly and freely as we should
  - Avoid controversial subjects

Push

ENGINEERS AUSTRALIA

- If we are unable to persuade the zombies to up their game through incentives, we need to consider ways to force them to change.
- Regulation of the profession for some may be the only option…

Where to from here

ENGINEERS AUSTRALIA

- As professional societies we need to;
  - Remain relevant to current membership
  - Become relevant to those in practice who could/should be members
  - Attract new members of the profession
  - Be seen, and be sought out as the voice of our industry
- If we don't have the numbers we cant be relevant

They really are everywhere…

ENGINEERS AUSTRALIA

ENGINEERS
AUSTRALIA

www.engineersaustralia.org.au

Balancing Process Safety and Personal Safety

Nathan Winter – Deputy Chairman, Safety Institute of Australia

Safety Institute of Australia Ltd

Certification Qualification Requirements

Certified "Generalist" OHS Practitioner

- OHS/WHS Diploma

Certified "Generalist" OHS Professional

- OHS/WHS Diploma
- OHS/WHS Degree
- OHS/WHS Graduate Diploma

Safety Institute of Australia Ltd

What is Process Safety?

What do you know about Process Safety?

Process Safety is a blend of engineering and management skills focused on preventing catastrophic accidents, particularly explosions, fires, and toxic releases, associated with the use of chemicals and petroleum products.

Source: AIChE – CCPS website http://www.aiche.org/ccps/about/process-safety-faqs#What%20is%20Process%20Safety

Safety Institute of Australia Ltd

Process Safety vs Personal Safety

"The fundamental difference between personal safety and process safety is that personal safety deals with low severity, high frequency risks, whereas process safety deals with high severity, low frequency risks. Process safety is not really 'safety' at all, and in fact is more akin to risk management."

Gary Pilkington

Source: IChemE website http://www.icheme.org/events/events/2013/safety-webinar-pilkington-20-september.aspx#.VNI7Ov4cSUI

Safety Institute of Australia Ltd

OHS Body of Knowledge

Chapter 31 "Risk"

Section 5.2 "Risk Management"

APPEA Process Safety Video

- Definition "Is all about the barriers that are put in place to keep our facilities safe and our products in the pipes, now these barriers can be people, process or plant related."
- Swiss Cheese Model
- Portrays safety specialist as someone that is only interested in PPE.
- I went away with the perspective that Process Safety was about "speaking up"

Hopkins Safety Indicators Article

Process safety indicators consist of;

- Fires
- Explosions
- Loss of containment events
- Deviations from safe operating limits

API's Recommend Practice 754: Process Safety Indicators for the Refining and Petrochemical Industries.

S.I.A. Safety Institute of Australia Ltd

Industry Associations and Driving Process Safety

- Generalist Safety Practitioners and Professionals
- Appropriately qualified
- Experienced
- Committed to Continuing Professional Development
- If desired "Certified"

S.I.A. Safety Institute of Australia Ltd

Code of Ethics

Victorian Code of Ethics and Minimum Service Standards for professional members of occupational health and safety (OHS) associations

1.3 Areas of competence

Perform work only in their areas of competence and within the limits of their abilities.

Safety Institute of Australia Ltd

SIA Code of Conduct

SIA Code of Conduct

- Members will give priority to the health, safety and welfare of the community ~ in accordance with accepted standards of moral and legal behaviour ~ during the performance of their professional duties;
- Members will perform their professional duties with integrity, honesty and equity while adhering to legal principles and operating within their area of competency;
- Members will not engage in any illegal or improper practices;
- Members words or deeds must not adversely affect the reputation of the Institute or the professional reputation of another person;
- Members will fulfil the terms and conditions of their employment or contract and avoid real or apparent conflicts of interest; and
- Members will continue their professional development and, therefore, the development of their profession.

Source: Safety Institute of Australia Ltd Website: http://www.sia.org.au/about/code-of-conduct/

Safety Institute of Australia Ltd

Summary

Safety Institute of Australia can help drive Process Safety by;

- Making Members aware of this specialist safety field.
- Reminding members to ensure that they only operate within their areas of competency.
- To engage Process Safety specialist advice in the event that it is required.
- Encouraging Officers and Mangers of applicable organisations to ensure that they engage and listen to the advice of Process Safety Specialists.

S.I.A. Safety Institute of Australia Ltd

Thank you for your attention

What Questions do you have?

S.I.A. Safety Institute of Australia Ltd

AUTHOR INDEX